Advanced Motorsport Engineering

Advanced Motorsport Engineering

Units for Study at Level 3

Andrew Livesey

LONDON AND NEW YORK

First published 2012
by Routledge
2 Park Square, Milton Park, Abingdon, Oxon OX14 4RN

Simultaneously published in the USA and Canada
by Routedge
711 Third Avenue, New York, NY 10017

Routledge is an imprint of the Taylor & Francis Group, an informa business

British Library Cataloguing in Publication Data
A catalogue record for this book is available from the British Library

Library of Congress Cataloging in Publication Data
A catalog record for this book has been requested

ISBN: 978-0-75-068908-3 (pbk)
ISBN: 978-0-08-094333-6 (ebk)

Typeset in Stone Serif
by RefineCatch Limited, Bungay, Suffolk

CONTENTS

AUTHOR'S FOREWORD

This is the second book to be specifically written for the formal training of motorsport engineers and follows on from the Level 2 book of the same series by the same author.

The motorsport industry in Great Britain is one of the biggest exporters of goods and services and its engineers are sought across the world. In the press you will read about Formula 1; underneath that pinnacle of speed and technology are 50,000 other racing drivers in the UK alone and to support them many more thousands of engineers, mechanics and service personnel.

Motorsport Engineering is an excellent subject to study as it teaches you to investigate and analyse situations, as well as learning a practical skill and investigating technology. Interestingly, the aerospace industry now looks to motorsport for new technology.

I hope that all of you who read this book and take up a career in Motorsport Engineering will have as much fun and success as I am having in my career.

William Andrew Livesey
MA CEng MIMechE FIMI
Andrew@BrooklandsGreen.Com
Kent
November 2011

ACKNOWLEDGEMENTS

Thanks to the following for their help and support with this book, either with information, direct or indirect support which enabled the writing process and my involvement in motorsport to continue.

Adrian Slaughter
Alan Flavell
Alex Heaton
Automotive Skills
Bernie Ecclestone
Bill Sisley
Brian Marshall
Brian Redman
Brooklands Museum
BrooklandsGreen.Com
Buckmore Park Kart Circuit
Bucks New Uni
Carol Dorman
Chris Pulman
Claire Kocks
Daniel Johnson
Dave Henderson (Rambo)
Evers Pearce
Express Garage, Herne Bay
Frank Groom
Greenwich University
Helix Autosport Ltd
Ian Goodwin
Institute of the Motor Industry
Jack Sealey Ltd
Jenson Button
John Hunt
John Ryan
John Surtees
Kevin Pilcher
Laser Tools
Linda Teuton
McLaren Cars
McLaren Racing
Mick Ellender

Mike Farnworth
Mike Shaw
Motorsport Academy
Motorsport Institute
Mr and Mrs Presgraves (Snr)
North West Kent College
Oxford Brookes University
Paul Presgraves
Paulio Racing
Performance Racing Industry (magazine)
Proton Cars (UK) Ltd
R T Quaife Engineering Ltd
Rallyspeed
Richard Davies
Seyed Edalatpour (Ali)
Shiva
Simon Hammond
Staffordshire University
Steven Swaffer
Terry Ormerod
The late James Hunt
Tool Connection
University for the Creative Arts
University of Brighton
University of Kent
Winston Sewell

Dedication

This book would not have been possible without the support and love provided by my family, that is my wife Jean and the 'kids' Rachel, Rebecca, Samuel, Steve, Andy, Gemma, Oliver, Pepper, Kye and Neve.

ABBREVIATIONS AND SYMBOLS

The abbreviations are generally defined by being written in full when the relevant technical term is first used in the book. In a very small number of cases, an abbreviation may be used for two separate purposes, usually because the general concept is the same, but the use of a superscript or subscript would be unnecessarily cumbersome; in these cases the definition should be clear from the context of the abbreviation. The units used are those of the internationally accepted *System International* (SI). However, because of the large American participation in motorsport, and the desire to retain the well known Imperial system of units by UK motorsport enthusiasts, where appropriate Imperial equivalents of SI units are given. Therefore, the following is intended to be useful for reference only and is neither exhaustive nor definitive.

α (alpha) angle – tyre slip angle
γ (lambda) angle of inclination; or measurement of air-fuel ratio
μ (mu) co-efficient of friction
ω (omega) rotational velocity
ρ (rho) air density
a acceleration
A area – frontal area of vehicle; or Ampere
ABS anti-lock braking system; or acrylonitrile butadiene styrene (a plastic)
AC alternating current
AED automatic enrichment device
AF across flats – bolt head size
AFFF aqueous film forming foam (fire fighting)
ATC automatic temperature control
bar atmospheric pressure
BATNEEC best available technique not enabling excessive cost
BCF bromochorodifluoromethane (fire fighting)
BS British Standard
BSI British Standards Institute
C Celsius; or Centigrade
CAD computer-aided design
CAE computer-aided engineering
CAM computer-aided manufacturing
C_D aerodynamic co-efficient of drag
CG centre of gravity
CI compression ignition
CIM computer integrated manufacturing
C_L aerodynamic co-efficient of lift
cm centimetre
cm^3 cubic centimeters – capacity; engine capacity also called cc. 1000 cc is 1 litre
CO carbon monoxide
CO_2 carbon dioxide
COSHH Control of Substances Hazardous to Health (Regulations)

CP centre of pressure
CR compression ratio
D diameter
d distance
dB decibel (noise measurement)
DC direct current
deg degree (angle or temperature)
dia. diameter
DTI dial test indicator
EC European Community
ECU electronic control unit
EFI electronic fuel injection
EPA Environmental Protection Act; or Environmental Protection Agency
EU European Union
F Fahrenheit
ft foot
ft/min feet per minute
FWD front-wheel drive
g gravity; or gram
gal gallon (USA gallon is 0.8 of UK gallon)
GRP glass reinforced plastic (glass fibre)
HASAWA Health and Safety at Work Act
HGV heavy goods vehicle (used also to mean LGV – large goods vehicle)
hp horse power (CV in French, PS in German)
HSE Health and Safety Executive; also health, safety and environment
HT high tension (ignition)
HVLP high volume low pressure (spray guns)
I inertia
ICE in-car entertainment
ID internal diameter
IFS independent front suspension
IMechE Institution of Mechanical Engineers
IMI Institute of the Motor Industry
in^3 cubic inches – measure of capacity; also cu in. Often called 'cubes' – 61 cu in is approximately 1 litre
IR infra red
IRS independent rear suspension
ISO International Standards Organization
k radius of gyration
kph kilometers per hour
KW kerb weight
l length
L wheelbase
LH left hand
LHD left hand drive
LHThd left-hand thread
LPG liquid petroleum gas
LT low tension (12 volt)
lumen light energy radiated per second per unit solid angle by a uniform point source of 1 candela intensity
lux unit of illumination equal to 1 lumen/m^2
M mass
MAG metal active gas (welding)
MAX maximum
MDRV mass driveable vehicle
MI Motorsport Institute
MIA Motorsport Industry Association
MIG metal inert gas (welding)

MIN minimum
MoT Ministry of Transport; also called DTp – Department of Transport and other terms depending on the flavour of the Government, such as the Department of the Environment Transport and the Regions (DETR), not to be confused with DOT which is the American equivalent
N Newton; or normal force
Nm Newton metre (torque)
No number
OD outside diameter
OL overall length
OW overall width
P power, pressure or effort
Part no part number
PPE Personal Protective Equipment
PSV public service vehicle (also used to mean PCV - public carrying vehicle in other words a bus)
pt pint (UK 20 fluid ounces, USA 16 fluid ounces)
PVA polyvinyl acetate
PVC polyvinyl chloride
r radius
R reaction
Ref reference
RH right hand
RHD right hand drive
rpm revolutions per minute; also RPM and rev/min
RTA Road Traffic Act
RWD rear-wheel drive
SAE Society of Automotive Engineers (USA)
SI spark ignition
std standard
STP standard temperature and pressure
TE tractive effort
TIG tungsten inert gas (welding)
TW track width
V velocity; or volt
VIN vehicle identification number
VOC volatile organic compounds
W weight
w width
WB wheel base
x longitudinal axis of vehicle or forward direction
y lateral direction (out of right side of vehicle)
z vertical direction relative to vehicle

Superscripts and subscripts are used to differentiate specific concepts.

SI Units

cm centimetre
K Kelvin (absolute temperature)
kg kilogram (approx. 2.25 lb)
km kilometre (approx. 0.625 mile or 1 mile is approx. 1.6 km)
kPa kilopascal (100 kPa is approx. 15 psi, that is atmospheric pressure of 1 bar)
kV kilovolt
kW kilowatt
l litre (approx. 1.7 pint)
l/100 km litres per 100 kilometres (fuel consumption)
m metre (approx. 39 inches)
mg milligram

ml millilitre
mm millimetre (1 inch is approx. 25 mm)
N Newton (unit of force)
Pa Pascal
µg microgram

Imperial Units

ft foot (= 12 inches)
hp horse power (33,000 ftlb/minute; approx. 746 Watt)
in inch (approx. 25 mm)
lb/in² pressure, sometimes written psi
lbft torque (10 lbft is approx. 13.5 Nm)

CHAPTER 1

Engine

Diagnose and Rectify Motorsport Vehicle Engine and Component Faults

This unit looks at the types of engines used in club level motorsport vehicles; it builds on the material covered in the Level 2 Motorsport Engineering book by the same author. It should be read in conjunction with the chapters on inspection, overhaul, and modification in this book.

IDENTIFICATION

Identification of the engine before working on it is very important. The VIN number will help identify the type, or classification, of the engine. The detail of the engine will be given in a separate engine number, the prefix will identify the engine type, and the serial number will identify the exact engine.

With motorsport engines the build may be completely different to standard, for this reason you should keep a log of the engine build, detailing all the components including part numbers, sizes, and any other variants.

ENGINE PERFORMANCE

The two common terms used in motorsport are:

Power – this is **work done** in unit time.

Torque – turning moment about a point.

For a deeper definition of these terms you need to start with work done. Work done is the amount of load carried multiplied by the distance travelled. The load is converted into force, the force needed to move the car for instance in Newtons (N). The distance is measured in metres. That is:

$$\text{Work done (Nm)} = \text{Force (N)} * \text{Distance (m)}$$

As we also express torque in Nm, so it is common to use the term joule (J).

RACER NOTE

Joule is a term for energy. 1 J = 1 Nm.

If we use a force of 10,000 N to take a dragster down a 200 m drag strip then we have exerted 2,000,000 Nm, or 2,000,000 J. We would say two mega joules (2 MJ). We would need to get this amount of energy out of the fuel that we were using.

The force is generated by the pressure of the burning gas on top of the piston multiplied by the area of the top of the piston. So the work done is mean (average) force of pushing the piston down the cylinder bore multiplied by the distance travelled.

Example

The work done during the power stroke of an engine where the stroke is 60 mm and the mean force is 5 kN

$$\begin{aligned}\text{Work Done} &= \text{Force} * \text{Distance}\\ &= 5\text{ kN} * 60\text{ mm}\\ &= 5{,}000\text{ N} * 0.06\text{ m}\\ &= 300\text{ J.}\end{aligned}$$

FIGURE 1.1
LOLA T70B V8 engine

RACER NOTE

The mathematical symbols are those found on your calculator or mobile phone, * is multiply and / is divide.

The same mean force is going to create the torque, this time we are going to use the crankshaft throw – this is half of the length of the stroke.

Example

Using the same engine

$$\begin{aligned}\text{Torque} &= \text{Force} * \text{Radius}\\ &= 5\text{ kN} * 30\text{ mm}\\ &= 5{,}000\text{ N} * 0.03\text{ m}\\ &= 150\text{ Nm.}\end{aligned}$$

The work done by a torque for one revolution is the mean force multiplied by the circumference. The circumference is $2\pi r$ so

$$\text{Work done} = F * 2\,\pi r$$

as $$Fr = T$$

so $$\text{Work done} = 2\,\pi T$$

That is for one revolution, for any number of revolutions, where n is any number, the formula will be:

$$\text{Work done in n revolutions} = 2\,\pi n T$$

Example

Using the same engine of the previous examples
The work done in 1 minute at 6,000 rpm will be:

$$\begin{aligned}\text{WD in n revolutions} &= 2\,\pi n T\\ &= 2 * \pi * 6{,}000 * 150\\ &= 5657\text{ kJ.}\end{aligned}$$

Power is, as we said, work done in unit time, which is:

$$\text{Power} = \text{Work Done/Time}$$

The motorsport industry uses a number of different units and standards for power, from our calculations we can use Watts (W) and kilowatts (kW) and then convert.

RACER NOTE

1 kW = 1,000 W.

$$1\text{ Watt} = 1\text{ J/second}$$

and $$1\text{ kW} = 1\text{ kJ/s.}$$

Example

Following on from our engine in the previous calculations and examples:

Power = Work Done/Time
= 5657 kJ/60
= 94.3 kW.

The term **Horse Power** (HP or hp) was derived by James Watt as the average power of a pit pony. These were small horses used to turn pulleys to draw water from Cornish tin mines (pits) before steam power became more popular. He equated the power of his steam engines to a number of these pit ponies. For our purposes 1 HP equals 33,000 ftlb/minute.

In French, horse power is cheval vapour (CV); in German it is Pferda Stracker (PS).

For conversion purposes 1 HP is equal to 746 W.

The two standards for measuring power are the SAE system (USA) and the DIN system (Continental Europe).

Volumetric efficiency – in other words the efficiency of the engine of getting in fuel and air to fill the cylinder. The fact that an engine has cylinders of 1 litre (61 cu in) does not mean that you are getting

FIGURE 1.2
JCB Diesel world land-speed record holder engine

that amount of air and fuel into the cylinders. The flow of gas is affected by a number of points, mainly:

- size, shape and number of valves
- valve timing
- size and shape of the inlet and exhaust ports
- shape and location of combustion chamber
- bore to stroke ratio
- engine speed
- type of induction tract – air filter, carburetter or throttle body, inlet manifold
- type of exhaust system – manifold, silencers, cat, and pipe layout.

The volumetric efficiency is calculated by measuring the amount of air entering the engine and finding this as a percentage of the actual volume of the engine. The formula is:

Volumetric Efficiency = Actual Air Flow/Capacity of Engine

The air flow can be measured on the test dynamometer by using either an air flow meter on the inlet or calculating the air flow from the pressure drop across an orifice. You would normally do this over a set time period so that the air flow is calculated in litres per minute (cu in/minute or cu ft/minute).

A typical engine might have a volumetric efficiency of 80%, a well tuned naturally aspirated engine may have an efficiency of 100% or even 105% with throttle bodies and all the right manifolds. And because the turbocharger is pressurising the air into the cylinder, the figure will be over 130%.

The greater the amount of air that can be packed into the cylinder, the more power the engine is going to give out. The saying is, ***there aint no substitute for cubes***. That is, the more cubic inches (litres) of air, the more power the engine will develop.

Thermal efficiency – from volumetric efficiency you saw that you need to get the engine efficient in getting the air (actually air and fuel) in to and out of the engine. That is one thing. Now when the air and fuel is in the engine it needs to be burnt efficiently to produce as much energy as possible to be turned into power and that conversion of burning gas into turning the crankshaft needs to be as efficient as possible to get the maximum power out of the engine.

Thermal Efficiency = Power Output/Energy Equivalent of Power Input.

The power output is easy to measure on a dynamometer, either as an engine in a test rig, or the complete vehicle on the rolling road.

RACER NOTE

Lots of drivers know the power output of their engines, very few know the efficiency of the same engine – greater efficiency means fewer refuelling stops and hence faster race times, in other words more wins.

To calculate the energy input you are going to need to know:

- Mass of air used – that is volume of air multiplied by the density. The density depends on altitude, temperature and weather conditions.
- Mass of fuel.
- Specific calorific value of the fuel – how much energy it produces for a given amount.

The calculation for the mass of air follows on from the measurements of air flow from the volumetric efficiency calculations. The calculations for density may be made by either sampling or using tables.

RACER NOTE

Many racing classes require the use of a control fuel – so all competitors are equal.

STP – Standard Temperature and Pressure

STP – Standard Temperature and Pressure – is defined as air at 0 °C (32 °F) and 1 bar (14.7 psi).

Also STP – commonly used in the Imperial and USA system of units – as air at 60 °F and 14.7 psi.

NTP – Normal Temperature and Pressure

NTP – Normal Temperature and Pressure – is defined as air at 20 °C (68 °F) and 1 bar (14.7 psi).

Air density at sea level at NTP is 1.2 kg/m^3 (0.075 lb/ft^3).

RACER NOTE

Have you ever noticed that on a cold, damp day the engine runs well and sounds well – this is because the air is at maximum density and the water cools the cylinder to prevent detonation.

The energy from fuel is discussed under fuel composition later in this chapter, examples of calculations can be found on the website.

RACER NOTE

Some simple changes, like chipping, may increase the power by 5% or 10% but increase fuel consumption by 40% – customers may have a shock at the increase in fuel costs.

Valve timing – is very important on a competition engine. The basic principles are to open the valves as quickly as possible, as great a distance as possible, and for as long as possible. The problem is that if you do this, the engine will need to run fast to maintain gas velocity, so it is likely to stall at low engine speeds, or at least produce low power. To this end many vehicles use variable valve timing.

If you fit a high performance camshaft you need to ensure that the timing is exact – especially on engines where fully open valves may touch the piston crown. To this end a **vernier adjustable timing pulley** may be needed.

Cam design – there are several types of cam design, these are: **constant velocity**, **constant acceleration** and **simple harmonic motion**.

COMBUSTION

Flame travel – the combustion of the air and fuel mixture is a burning action – we often refer to it as an explosion because it is very fast, however the fuel burns with a flame which is travelling very quickly. At 6,000 rpm each cylinder fires 50 times each second; in a Formula 1 engine at 22,000 rpm this is 183 times per second. In this time the piston has completed four strokes, so the stroke time is a quarter of this, that is the maximum duration of the burning. If it were between TDC and BDC, it would be 0.005 s (5 ms) at 6,000 rpm and 0.0016 s (1.6 ms) at 22,000 rpm. On race engines the spark plug needs to be as near to the centre of the combustion chamber as possible to give even burning of the gas. The faster the engine speed is the more important this is. Flame travel should not be confused with spark plug burn time which is the length of time that the plug is sparking.

Pre-ignition, **Post-ignition**, **Pinking** and **Detonation** are often confused as their symptoms and effects are almost identical; that is the air and fuel are burnt in a noisy manner, usually a knocking noise, and there is a loss of power or uneven running. **Running-on** is associated with these symptoms too. Pre-ignition is when the fuel is ignited before the spark occurs, this is usually by something burning, or overheating in the combustion chamber. Often this is the spark plug insulator, a carbon deposit or a section of damaged cylinder head gasket. Post-ignition is when the fuel burns late, or more likely combustion continues after the engine is switched off – the engine continues to run for a while when the ignition is switched off, this is running-on. Pinking, also called **knock**, is the noise made by the mixture burning too quickly, or on two flame fronts, caused by the ignition timing being too advanced or too low octane rating fuel being used. Detonation is when pockets of fuel burn in an irregular way, usually because of poor air fuel mixing, incorrect ignition timing or poor combustion chamber design.

Octane rating – there are two methods of measuring fuel anti-knock ability, these are:

- **Research Octane Number** (RON), a measure of anti-knock during acceleration under medium load.
- **Motor Octane Number** (MON), a measure of anti-knock during acceleration under heavy load.

As neither are perfect measures, an average of the two is often used; this is called the **Anti-knock Index**

Anti-knock Index = RON + MON/2

Cetane rating – the diesel equivalent of octane rating.

THINK SAFE

Petrol is highly flammable – race petrol is especially volatile.

CYLINDER HEAD

Valve layout – the valve layout for high performance engines is designed to get the maximum amount of air and fuel into the engine, and then out again as exhaust gas. The most common design uses four valves, the inlet valves being larger than the exhaust valves as the exhaust gas is forced out by the piston under high pressure. The gas flows from one side of the engine to the other to reduce the time taken; this allows an increase in engine speed.

Cooling – for good cooling aluminium cylinder heads are almost universal – they dissipate the heat very quickly – coolant (water) jackets may be opened out to improve coolant flow.

Valves – the use of a variety of materials is used to give longer valve life and better cooling. Typically high chromium alloy steel to resist corrosion and give wear resistance. Valve head shapes tend to be with thin heads – often called penny on a stick – for minimum weight and less mass for heat build up. Variations of seat angles and seat materials are used to give good sealing.

RACER NOTE

The thinner the valve seat the tighter will be the seal; conversely, the wider the seat the better the heat transfer from valve to cylinder head.

With aluminium alloy cylinder heads seat inserts are needed to give the required hardness, iron cylinder heads may also have inserts for competition use – these are often fitted at the same time that the cylinder head is machined to fit larger valves. Valve seat inserts may be screwed – now rare – or press fit into the cylinder head.

To improve thermal efficiency, two factors of combustion chamber design should be considered, these are: **Swirl Ratio** and **Surface to Volume Ratio**, the formulas are quite self explanatory.
They are:

Swirl Ratio = air rotation speed/crankshaft speed

The higher this figure is the better the engine will run – look for figures over 5.

Surface to Volume Ratio = surface area of combustion chamber/volume of combustion chamber

The lower this number is the better the engine will run – a sphere will give the best ratio.

Valve springs – double or triple valve springs may be used to enable high revs.

Camshaft – when building a competition cylinder head you should pay attention to the fit of the camshaft(s). If possible use white metal bearings and line bore to ensure perfect straightness.

RACER NOTE

Turn the camshaft like the crankshaft when setting it up; that is ensure that it turns freely at each step of tightening the caps, or when inserted – you will need to do this before fitting the valves.

Standard camshafts tend to be cast iron or low grade steel – for competition use you will need a high grade alloy steel. These can be hardened by Nitriding – lower quality competition camshafts may be tockle hardened, Tufftrided, or simply coated to give a better wearing surface.

Camshaft drive – for competition use high tensile bolts will be used.

Gas flow – this is a commonly misused term, or used as a catch all for anything to do with cleaning and polishing cylinder heads – gas flowing. The volume and velocity of gas flow through a cylinder head is measured on a **flow bench**. The cylinder head is attached to the flow bench so that the air flowing through it can be measured. The exhaust valve is kept firmly shut whilst the inlet valve is opened in small increments and the flow recorded. The same can then be done with the exhaust valve. The modifications are then made to the relevant part of the cylinder head – typically the inlet port is smoothed and polished, then the flow is measured again. It is not the absolute flow rate which is important – this may be difficult to measure in a finite way – the percentage improvement is the indicator looked for. A 5% improvement is very good. The factors effecting gas flow are:

- port shape
- port finish
- manifold to port fit – do they align?
- gasket fit – does it overlap the port?
- valve seat angle
- valve size
- valve opening.

SHORT BLOCK ASSEMBLY

Materials – for lightness and cooling dissipation aluminium alloy is the obvious choice – but this will depend on the racing class. To improve aluminium block strength and improve wear resistance **cryogenic treatment** is often carried out – check the regulations to see if it is allowed for the class. This involves taking the temperature of the block down to a temperature of about –180 degrees C (–300 degrees F) for a period of about 48 hours. This smoothes out the grain structure of the metal, making it stronger.

RACER NOTE

You can cryogenically treat any part of the car – brake discs (rotors) and pads are a common and economical choice.

Liners are used in aluminium blocks to give the essential resistance to wear. They may be **dry liners**, or **wet liners**. Wet liners are the most common to give easy to make open blocks. If you are cryogenically treating the liners you will need to bore them out about 0.002 in (two thou) to return them to true round – you must bear this in mind if you are blueprinting the engine.

RACER NOTE

Never hit liners, even with something soft – they easily distort.

Drillings – when building a race engine check the drillings and passage ways to ensure that they are clear and the correct size to give sufficient water and oil flow.

Crankshaft – made from high grade alloy steel with a high nickel content so that it can be nitrided, this involves leaving it in a bath of ammonia at 500 degrees C (900 degrees F) for about 24 hours.

When fitting the crankshaft ensure that it turns after each bearing cap is tightened. A very small amount of very light (5 SAE) oil should be placed on each journal and bearing shell; but ensure that there is no oil behind the shell. Also check for crankshaft end float after fitting the thrust bearings. End float must be the minimum to ensure con-rod alignment and clean clutch operation – remember the clutch thrusts against the crankshaft thrust bearings.

Con-rods – increasingly made from carbon fibre composite as this method allows relatively low cost batch production. The alternatives are high strength aluminium and titanium. Consider the speed at which they are travelling and you will appreciate the need for aerodynamics in their design. Changing the engine stroke or the piston height may mean changing the con-rods.

Bore-stroke ratio – this is simply the ratio of the two dimensions = bore/stroke (mm or inches).

If the bore is larger than the stroke, the engine is said to be an over square engine; if bore and stroke are the same it is said to be a square engine; if the stroke is greater than the bore then it is called a long stroke engine. Generally race engines are over square and rev at high speeds, this can be over 20,000 rpm. Older engines tend to be long stroke, running at a lower speed and producing more torque than ultimate power.

Balancing – pound (dollar) per horse power, balancing an engine gives the best return. Not only does it ensure smooth operation, and longer life, it also allows higher revs at the top end. Balancing breaks down into two parts, these are **static** and **dynamic**.

Static balancing can be done on the bench with a simple weigh scale. Dynamic balancing needs a specialised machine. To carry out static balancing you start with the pistons. Weigh each piston noting its weight – now remove material from the heavier ones until they all weigh the same. Reduce the weight of the heavy pistons by removing material from the bottom inside of the skirt or a relatively unstressed part of the boss.

RACER NOTE

If you have access to more than one box of pistons of the same size weigh them all – you may find the number that you want that are all the same to save machining.

Next are the con-rods. Support the little end on a piece of wire (or similar) and weigh the big ends on the scale. Do the same supporting the big end. Remove metal by drilling small holes until they are all

FIGURE 1.3
Lightened flywheels

the same weight. Weight can be added on steel rods by drilling small holes and filling them with lead – make sure that they are very clean and use flux so that there is no risk of the filler coming out.

You can statically balance the flywheel and clutch by assembling them onto the crankshaft, supporting the crankshaft in vee blocks and spinning, then wait until it stops. The heavy part will be at the bottom. Again remove weight by drilling small holes in unstressed parts. When it is balanced you should be able to stop the assembly at any point without it rotating.

Properly statically balanced – dynamic balance should be easy – indeed if on a budget, static balance only will give very good results.

Surface finish – surface finish is important for mating surfaces and surfaces which are subject to air or other fluids flowing or over them. On a race engine, good surface finish makes the the engine look good – a very important point. Nicely polished parts make the engine attractive.

Blue printing – this is another often misused term. It simply means building the engine to the manufacturer's specification – in other words the blue print (or drawing – as CAD means that we no longer print drawings in blue). When blue printing you should pay detailed attention to each part fit, and ensure that it is the correct size within design tolerance.

INDUCTION

Gas laws – air is not always what you think it is. You may think that air goes through the engine like sausages through a sausage machine. You may think that a chunk of it goes into the cylinder to be processed, and then it comes out as exhaust gas. Certainly air is mixed with petrol; it does go into the cylinder; it is burnt and it does come out as exhaust. However the gas is travelling at several hundred miles per hour (hundreds of kilometres per hour). And it travels in waves, not in sausage-like lumps.

Induction falls into three categories, these are:

- Open induction, usually the carburetter, or the throttle body has an open ram pipe.
- Closed induction, this uses a plenum chamber with air filter.
- Forced induction, as with a turbocharger, or supercharger.

The ram pipe length can be changed to suit the engine operating speed – this is very much a trial and error activity. The plenum (Latin for full) chamber is designed to give the engine a supply of still air. With forced induction air is in effect pumped into the cylinder under pressure.

The ram pipes are guiding the air into the cylinder, and helping to control the waves.

RACER NOTE

Changing from plenum chamber to ram pipes can add 5% to the power output of some engines.

EXHAUST

Exhaust gas travels in the same way – that is it comes out in waves; the waves are generated by the piston, in the same but opposite way to the induction waves. The piston going up the bore pushes the gas out, not as a sausage, but as a pulse. The pulse velocity depends on a number of factors which we will look at later in this section. First of all let us look at the speed of sound – exhausts as you know make a sound, so the pulse wave is travelling at the speed of sound. The speed of sound changes with a number of variables.

HISTORIC RACER NOTE

Race cars and bikes at Brooklands had to have their exhaust tail pipe extending past the rear wheel spindle, to satisfy the complaining residents of near by St Georges Hill – this led to fancy fish tail silencers.

The **speed of sound** in air (C) is calculated from the heat capacity (γ) of air (1.4) the air pressure (p) and the air density (ρ)

$$C = \sqrt{\gamma\, p/\rho}$$

Example

At NTP

$$C = \sqrt{1.4 * 101000/1.2} = 343 \text{ m/s}$$

The speed of sound varies with temperature (T) in the ratio:

$$C2/C1 = \sqrt{T2/T1}$$

Taking T1 as NTP of 20 degrees Celsius and C1 as 343 m/s and remembering that this should all be in absolute temperature form Kelvin (K). At 800 degrees C the speed of sound will be:

$$C2/343 = \sqrt{800 + 273/273 + 20}$$
$$= 656 \text{ m/s}$$

Now we have the speed of sound in the very hot exhaust. This allows us to work out the length of the exhaust **primary pipe** to suit the engine speed. You'll see lots of variations of exhaust manifolds and systems. The primary pipe length from the exhaust valve to the 'Y' joint into the rest of the system; or the open pipe on a dragster is important to give maximum power at any particular speed. The reason for this is that as the exhaust valve opens it sends a pulse wave at the speed of sound – we calculated that – down the primary pipe. When it gets to the end of the primary pipe and the gas can expand, the sound wave travels back down the pipe to the exhaust valve, this then returns back to the end of the primary pipe.

TRY THIS

Think of the exhaust gas as a wave on the sea shore – it goes in and out even though the tide is coming in all the while.

If you can get the primary pipe to be the length that will be just right for gas to go up and down between the exhaust valve opening and closing it will leave a negative pressure (vacuum) at the exhaust valve as it closes, this will scavenge (clean out) the exhaust gas from the engine most efficiently. So, let's look at how we do this. We know that time taken to cover a distance is that distance divided by the velocity:

$$\text{Time (t)} = \text{distance (d)/velocity (v)}$$

The distance is from the exhaust valve to the end of the primary pipe and back, we'll call this 2L. The velocity is the speed of sound (C).

The time varies with engine speed and valve period (the number of degrees the exhaust valve is open, for example 120 degrees). 120 degrees is 0.333 of a revolution (360 degrees). At 6,000 rpm equates to 100 rev/sec, so 0.333 of a revolution will take 0.00333 seconds, that is: 0.00333/100 = 0.00333. We calculated that the speed of sound at 800 degrees C is 656 m/s. So the distance travelled by the sound wave at 656 m/s in 0.00333 s is:

$$0.00333 * 656 = 2.18 \text{ m}$$

The length of the primary pipe is half of this, remember it is down and back, so it is 1.09 m.

TRY THIS

Calculate the length of the primary pipe if the engine speed is 10,000 rpm.

HISTORIC RACER NOTE

The same principle applies to the induction, which is why you see old racing cars with long inlet manifolds and long ram pipes.

Helmholtz Theory – if you look at the inner tube in a silencer you see that it is drilled with lots of holes. Helmholtz worked out that if you chop the end off a wave it will reduce its energy, in other words the noise which it makes. So, as the sound wave passes each hole in a silencer, a part of it goes through the hole reducing the energy and hence the noise.

THINK SAFE

Exhaust gas is both hot and poisonous.

CARBURETTERS

On race cars carburetters are still popular even though we have not had them on road cars since the 1990s. The most popular is probably the **SU**. The initials are an abbreviation for Skinner's Union. British Army Colonel George Herbert Skinner designed this carburetter and developed it for his daughter who used to race an Austin 7 at Brooklands. The SU is a constant depression (CD) variable choke carburetter. The principal of operation is that as the volume of air flowing through the carburetter increases the choke size increases. This ensures a constant depression in the choke venturi to draw fuel in through a jet tube with a tapered needle. The shape of the taper on the needle controls the flow of fuel. By altering the needle diameter the mixture strength is varied. The carburetters come in a variety of sizes which are related to the maximum choke diameter in inches. The popular sizes are HS2 which is $1^1/_4$ inch and HS4 which is $1^1/_2$ inch. **Stromberg** made a similar carburetter; the main difference is that the Stromberg uses a rubber diaphragm in place of the metal piston of the SU. The problem with the Stromberg is that if the rubber diaphragm punctures the carburetter does not work. The equivalent Stromberg sizes are CD 125 and CD 150, which is 1.25 inch and 1.50 inch.

The other race carburetter, which is still manufactured, is the **Weber**. Weber was a colleague of Ferrari and Weber Carburetters became revered because of their use on Ferraris, and then became popular and readily available because of their adoption by Ford and Lotus for vehicles such as Ford's Lotus Cortina, the forerunner of the current Ford ST series. The Weber's major claim is that it is a die-cast body with every part screwed to it separately. There are three major design types, these are:

- **DCOE** – horizontal twin choke
- **IDA** – vertical twin choke
- **DCD** – vertical progressive twin choke for smaller engines.

The chokes come in a variety of sizes, referred to by their diameter in millimetres, 40, 45 and 48 being the most common for DCOE and IDA; 32/36 being the DCD norm.

The thing about the Weber is that you can change every part to alter its operating characteristics over the engine speed range. The normal procedure is to have the car on the rolling road and change jets to observe power changes or torque characteristics. However, before you do this, Webers are not cheap and a small box of jets can cost you the same amount of money as a new small car.

For large American V8 engines, you will find **Holley** and **Edelbrock** carburetters are the most common. These are 4-barrel carburetters, that is, they have four choke tubes. For a V8 engine you would normally use two 4-barrels, so having one choke to one cylinder. USA carburetters are normally rated by the amount of air that can flow through them in a minute. For racing purposes Holley and Edelbrock produce a range of carburetter sizes, with 500 CFM – that is 500 cubic feet of air per minute – being the most popular size. The Edelbrock carburetters feature an **electric choke** so that you do not need a throttle cable to the accelerator pedal (USA – gas pedal).

A quirky carburetter found on both racing cars and motor cycles is the Scottish **Minnow Fish**; this is a precursor to the single point fuel injection in that it used the pressure of the fuel from the electric fuel

FIGURE 1.4
Relocated fuel pump on racing BMW

pump to force petrol into the choke tube (venturi) instead of relying on the depression from the venturi effect.

Also found on small car engines are the sliding valve **Amal** and other motorcycle carburetters. Motorcycle carburetters work like SUs; but

FIGURE 1.5
Relocated fuel pump on racing BMW (close view)

instead of the air flow raising the piston as it does in the SU and the air flow being controlled by the throttle butterfly, the motorcycle carburetter has a slide to which the needle is attached. The cable raises the slide and needle at the same time, this controls both the air flow and the petrol–air mixture strength.

PRESSURE CHARGING

Pressure charging is about trying to get as much petrol and air into the cylinder as possible, pumping it in to fill it up. The higher the pressure when the piston is at BDC the higher the pressure the mixture will burn at. The compression ratio is limited by the physical factors of the block and head – fitting in the valves and the pistons. In some instances, the pressure charged version of an engine may have a lower compression ratio than the naturally aspirated version.

RACER NOTE

Naturally aspirated means no pressure charging.

However, we should bear in mind that the power output of the engine depends on the **mass of petrol and air** that we burn.

TRY THIS

The mass remains the same, but when we increase the temperature either the volume or the pressure increases, remember Boyle's and Charles' Combined Laws.

So, we charge the cylinder with air and fuel at the highest pressure and the coolest temperature we can. We keep it cool by using a cool pickup point for the air entry at the front of the vehicle – this is also the highest pressure point before the aerodynamic part of the body. Then we place an inter cooler between the turbocharger and the inlet manifold.

The pressure which we are usually concerned with is the **boost pressure**; this is the pressure above atmospheric pressure which is going into the engine. This is the pressure which you can read on the boost gauge if one is fitted. For calculation purposes you would need to add the boost pressure to atmospheric pressure to get the absolute pressure. So when at high altitude the boost pressure would need to be higher for the same effect.

If the pressure charger is allowed to produce too much pressure, the engine could be damaged, so a variety of pressure limiting devices are used.

Boost pressure may vary between 0.3 bar (5 psi) on an old race car to 3 bar (45 psi), or even more in extreme applications. Pressure charging increases the power output by massive percentages – 30% being easily attainable.

TRY THIS

Absolute Pressure = Atmospheric Pressure + Gauge Pressure.

Superchargers (also called blowers) are driven by belt from the crankshaft. In some cases a magnetic clutch is fitted to the supercharger drive pulley so that drive may be engaged and disengaged to driving conditions. The advantages of a supercharger are:

- Immediate response to increase in engine speed.
- Straight line power increase – power increase is almost directly proportional to engine speed.
- Driven directly from the crankshaft via a belt.
- Can be controlled by an electronic clutch.
- Maximum pressure can be controlled by an electronically operated valve.

The disadvantages of the supercharger are:

- Got to be mounted in line with the engine crankshaft.
- Uses power from the engine crankshaft to drive it.
- Is usually quite bulky.

THINK SAFE

Always allow a supercharged engine time for oil circulation to get to the supercharger and only switch off when the engine is idling or you could damage the supercharger and its drive belts.

HISTORIC RACER NOTE

Many old racers used cabin blowers – blowers used in old aircraft to maintain the air pressure for the pilot and passengers – as superchargers.

As a supercharger is a form of pump, the output will depend on the pumping volume, in other word the size of the supercharger. The skill is to ensure that the drive ratio of the supercharger matches the engines volumetric requirements. Changes in operating pulley sizes may be made to alter the operating characteristics. That is to make the supercharger turn faster, or slower, in relation to the engine speed.

TRY THIS

Look at a supercharger on a vehicle and work out the gear ratio between the crankshaft pulley and supercharger pulley – just measure their respective diameters. 1.5 to 1 is a typical figure, how does yours compare?

Turbochargers are driven by the exhaust gas, this is in effect using waste heat energy, unlike the supercharger which takes power from the engine crankshaft. The advantages of the turbocharger are:

- uses waste exhaust energy
- small and compact
- located in exhaust, so many options on position are available.

Because the operation is dependent on the exhaust gas, the operation cannot function until there is exhaust gas turning the turbine side. This means that there is a delay period between pressing the accelerator and the turbocharger cutting in to boost the induction pressure, this is called **turbo lag.**

THINK SAFE

When starting a turbo engined car always do so on no throttle and allow it time to warm up. Always let the engine idle for a few seconds before switching off to prevent the turbocharger from running without lubricant.

Turbo lag can be reduced by:

- using the correct size of turbocharger for the vehicle
- using the correct A/R ratio of turbocharger
- using the most suitable boost control system
- chipping for a rich fuel mixture.

The disadvantages of the turbocharger are:

- turbo lag
- high operating temperature – over 800 degrees C (1,500 degrees F)
- need for special lubricants.

The **size of the turbocharger** depends on the application; big does not always mean best. A small turbocharger running fast may pump the same amount of air as a large one running slow. The smaller one will react more quickly to changes in engine speed.

A/R ratio – this is the ratio of the cross-sectional area of the inlet tract – where exhaust gas goes into the turbocharger divided by the radius of the centre of the turbine wheel to the centre of this inlet tract. The A/R ratio is often shown on the side of the casing; typically values are between 0.5 and 0.9.

- A large A/R ratio gives a lower spin speed.

Boost control on a turbocharger can be by:

- Dump valve – this allows excess inlet pressure to go straight to the atmosphere.
- Blow-off valve – this balances the pressure between each side of the turbine.
- Waste-gate – this allows the exhaust gas to by-pass the turbine and go straight to the exhaust, it is controlled by boost pressure.

AUXILLIARY SYSTEMS

Cooling – competition engines are usually run at full power, so the amount of heat generated is high.

RACER NOTE

Typically heat energy from burning the fuel goes:

40% power output
30% exhaust gas
30% cooling system.

To cope with the large amounts of heat which is generated, think a German Touring Car, or Big Block Chevy, developing say 750 BHP. That will be the equivalent of 562 BHP (30% of the heat) going into the cooling system. That equates to 420 kW (562 BHP times 746) – imagine the heat from 420 electric fires. This means that the typical competition engine's cooling system will probably feature some, if not all of the following:

- high capacity coolant (water) pump
- coolant with wetness additive
- high pressure hoses – with high hoop strength to prevent expansion
- high fixing pressure clips
- high capacity radiator – to hold more coolant
- high capacity electric fan – to move more air mass
- high flow coolant jacket, and elbows
- aerodynamic design for air intakes and air egress – the aerodynamics of cooling air flow is a complex subject
- highly rated drive belts for all under bonnet applications to cope with the high temperature.

FIGURE 1.6
Racing Audi under-bonnet view

The coolant pressure is run very high to maintain the coolant below boiling point, internal temperatures are in excess of 180 degrees C.

Lubrication – if the motorsport vehicle is using a **wet sump** system it will probably have an oil cooler, the oil cooler may be controlled by a thermostat. For maximum engine efficiency and longevity oil temperature should be kept below 150 degrees C (300 degrees F), however synthetic lubricants are capable of running at much higher temperatures. An oil temperature gauge is a useful fitment.

To give more effective cooling and allow the lowering of the engine by reducing the depth of the sump, which in turn lowers the centre of gravity (CG), a **dry sump** system is used. Dry sump systems are universal on single-seat cars. The scavenge pump picks up the lubricating oil from the sump and takes it to a reservoir. The pressure pump draws oil from the reservoir (or oil tank) and sends it to the engine's main gallery for distribution to the bearings, from where it returns to the sump.

Exercise

When you see an engine for the first time it is important to identify it correctly, for its service needs, this is sometime quite difficult – it gets easier with experience. Try this exercise (Table 1.1) to help you get the essentials. You may need to use the workshop manual, engine log, or ask your tutor, you may wish to complete this on a separate sheet. The more detail you add, the more you will learn.

TABLE 1.1

Question	Answer	Comment
Make		Could be specially made
Model		May be a special version
Capacity		Cubic centimetres, litres, or cubic inches
Bore and stroke		Has it been bored out or fitted with a different crankshaft?
Valve gear		State cam position and operating mechanism
Valve timing		IVO-IVC-EVO-EVC
Induction system		Type of set-up and manifolds
Exhaust system		Make and type of manifold, system and if a cat is used
Supercharger		Include drive ratio and boost pressure
Turbocharger		Include A/R ratio and boost pressure and control system
Ignition system		Include spark plug type and settings
Cooling system		Include top-up details, pressure cap rating and coolant type
Lubrication system		Wet-sump or dry-sump, lubricant rating and operating pressure and temperature

FAULT FINDING

Before you start fault finding on an engine the first thing to do is to consider the engine type and apply this to the basic operating principles. It does not matter what type of make the engine is, it operates on the same principles. So, from Level 1 you learnt the operating principles of 4-stroke petrol, 4-stroke diesel, 2-stroke petrol and 2-stroke diesel. If you know what type of operating system is used, and the engine layout, then you can find the fault. Most race car engines are 4-stroke petrol, so think induction, compression, power and exhaust. So if the engine will not start, you need to think what are the most likely causes of the problem, let's look at the possibilities (Table 1.2).

TABLE 1.2

No.	Check	Yes	No	Comments
1	Is the engine turning over fast enough to initiate combustion?	Next step	Is the battery flat, is there a bad connection, or is there another reason	Remember race cars often use slave batteries
2	Is there a spark to the plugs?	Next step	Possible ignition fault – trace circuit back to battery	Follow manufacturer's instruction with electronic
3	Is fuel getting to the combustion chamber?	Next step	Is there fuel in the tank? Check fuel lines and pump for operation	Remember petrol is highly flammable
4	Check that the cam belt is not broken			You probably noticed this when you were checking the spark

RACER NOTE

When fault finding the following points are worth noting:

- The AA and RAC report that the most common reason for car non-starting is battery failure.
- Simple ignition faults and lack of fuel account for most road vehicle breakdowns.
- The cam belt usually breaks when an engine is being accelerated following a period of idling – generally it is advisable to replace them every time you need to work in that area of the engine.
- On a race car the fault is likely to be the one you never thought of!

CHAPTER 2

Chassis

Diagnose and Rectify Motorsport Vehicle Chassis System Faults

This chapter looks at the rolling chassis; that is the chassis systems and units of the vehicle without the power unit and transmission. It builds on material covered at Level 2.

VEHICLE IDENTIFICATION

It is wise to always check that the vehicle is what you think it is. The first stop is the **vehicle identification number** (VIN). The one shown in Figure 2.1 is for a Peugeot 307 following the European Union (EU) standard.

There are other standards in use, all very similar. These are the ISO standard and the USA standard developed by the Society of Automotive Engineers (SAE).

The standards are based on 17 digits. The first three digits identify the country and the manufacturer. The forth to ninth digits are the Vehicle Identifier Section (VIS). The tenth to seventeenth digits are the Vehicle Identifier Section (VIS) – or serial number.

With the USA system the ninth digit is a check digit – that it is a digit calculated from a formula based on the other digits in the sequence. So if some body changed one of the digits, for instance the engine capacity, they would also need to change the check digit. This is to prevent car crime.

Before the introduction of the VIN, and in some cases overlapping this, you will find **chassis numbers** and separate **body numbers** too. On formula cars you will find a chassis number or a **serial number**. Motorcycles will be found with a **frame number**.

You will also need to exercise some care in verifying the various numbers when dealing with certain cars. **Heritage organisations** supply certificates to prove genuine certain documented vehicles.

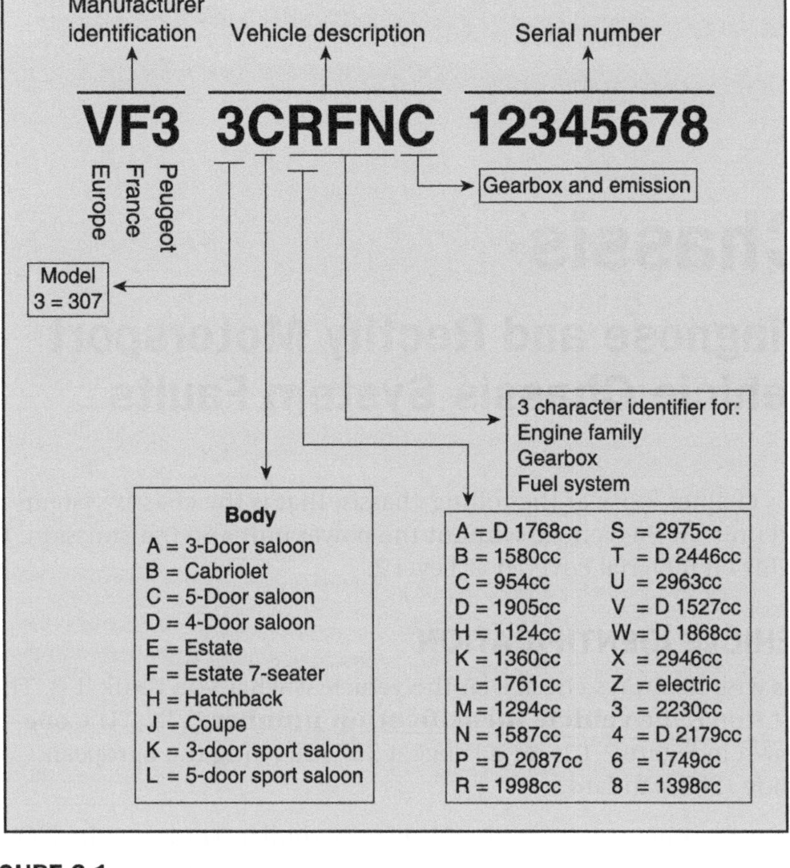

FIGURE 2.1
VIN identification for a Peugeot 307

Companies such as British Motor Heritage Ltd produce complete body/chassis units for many specialist vehicles such as the ever popular MG and Mini, these incorporate improvements of design and manufacture.

TRY THIS

Look at the VIN plate on any vehicle and write down its identity.

STEERING AND SUSPENSION TERMINOLOGY

This section sets out to describe and define a range of suspension terms and nomenclature relevant to the design of motorsport vehicle suspension.

Anti-dive and anti-squat – anti-dive is achieved by inclining the upper and lower wishbones so that their axes intersect at a point behind the suspension. This gives the equivalent of a leading arm suspension so that when the brakes are applied the braking force is applied in the anti-clockwise direction which lifts the front suspension against the weight transfer which is trying to depress the suspension in the opposite direction. The rear suspension is treated in the same way but the angle is reversed so that the point of intersection is in front of the suspension.

Axle tramp – is the sideways movement of the axle relative to the chassis which can occur on live-axle cars when cornering under power. Often this takes the form of an oscillation as the free play is taken up in the suspension joints. It can also combine with wheel patter to cause the loss of traction.

Camber and swivel axis inclination – on traditional suspension systems the camber of the wheel and the swivel axis inclination (also called king pin inclination – KPI) were designed to intersect at the road level, called centre point steering. Most current systems have negative scrub radius. That is the intersection of the camber and swivel pin, axes is above the road surface. The reason for this is, that with centre point steering, if a front tyre blew out there would be a steering torque about the steering axis. The intersection point would no longer be at a centre point, but instead would be below the road surface, the rolling resistance of the wheel would generate a torque about the steering axis. With negative scrub radius the intersection point is above the road surface so that in the case of a tyre failure the change in the intersection point would default to centre point steering. This prevents the vehicle veering to one side of the road or another in the case of a tyre becoming deflated. The distance between the tyre contact patch and the point of intersection between the steering angle and the road determines the amount of torque steer.

Variations in the camber angle can be brought about by changes in the swivel pin position. With modern multi-link systems this variation is very complicated; a method of analysis has been developed at the Hyundai Motor Co.

Castor angle – is the amount of trail between the wheel contact patch and the point where the steering axis intersects the road, this gives the steering wheels their self-aligning torque, so that the wheels return to centre after a corner and maintain stability at high speed on straight roads.

On a level road the torque is a function of the castor angle, that is the rearward inclination of the (imaginary) swivel pin ($\angle_c$), usually between 1 and 7 degrees. Large values of castor angle are sometimes used to give stability and high self-aligning torque; this is quite effective but can have an adverse effect on the handling. When a vehicle is heavily laden, such as with the addition of rear seat passengers, or a boot full of luggage, the rear suspension is usually the more compliant. The dipping of the body at the rear as the springs are compressed will increase the castor angle, this will, in turn make the steering heavier. The swivel pin inclination angle ($\angle_{kp}$), that is the angle by which the swivel pin (imaginary) leans transversely inwards at the top to give either centre point steering or a negative scrub radius, can also affect the amount of castor. That is, during cornering the swivel pin inclination may move because of two reasons. One is the movement of the theoretical centre of the pivot during steering; the other is the lateral forces which alter the shape and the position of the tyre contact patch. If the centre of pressure (CP) is in front of the centre of gravity (CG) then a side wind will exert a lateral force on the steering which will cause the steering to be turned in the same direction as the wind pressure. This can be overcome by fitting tail fins which move the CP rearwards, or the forward positioning of the engine and gearbox to move the CG. An alternative is to design in negative castor, but the steering will feel somewhat twitchy. The 1934 Auto Union 16-cylinder racing car is a good example of extreme positioning of the CG which gave such a variation in handling characteristics that the only driver to control the vehicle was Hans Stuck who broke seven speed records with her in the first season of competition.

Co-ordinate systems – vehicle movements and motions can be described in terms of co-ordinate systems; there are two systems, one which relates directly to the vehicle, the other which relates the vehicle to the earth – or its travel on the road.

FIGURE 2.2
Adjustable top link

VEHICLE FIXED CO-ORDINATE SYSTEM

The vehicle motions are defined with reference to a right-hand orthogonal co-ordinate system by SAE conventions. The co-ordinates originate at the centre of gravity (CG) and travel with the vehicle, the co-ordinates are:

x – forward and on the longitudinal plane of symmetry
y – lateral out the right side of the vehicle
z – downward with respect to the vehicle
p – roll about the x axis
q – pitch about the y axis
r – yaw about the z axis.

EARTH FIXED CO-ORDINATE SYSTEM

Vehicle attitude and trajectory through the course of a manoeuvre are defined with respect to a right-handed orthogonal axis system fixed on the earth. It is normally selected to coincide with the vehicle fixed co-ordinate system at the point where the manoeuvre is started, the co-ordinates are:

X – forward travel
Y – travel to the right
Z – vertical travel, positive is downwards
ψ – heading angle
ν – course angle
β – sideslip angle.

A similar co-ordinate system has been devised by the SAE for vehicle tyres.

DRIVING CHARACTERISTICS

THINK SAFE

Only test drive on the track on test days.

Driving and braking forces – the suspension must transmit between the chassis and the road the driving and braking forces. The rolling radius of the road wheel affects the amount of tractive effort (TE) which can be transmitted. At this point the coefficient of friction between the tyres and the road should also be considered. Race car drivers enjoy being able to spin the wheels, so this must be taken into account. Indeed, wheel spinning is often used to increase the coefficient of friction (μ), on dragsters the tyres are 'lighted' by spinning the wheels so that they get hot and momentarily set on fire to melt the outer rubber. The suspension mounting points must be capable of transmitting large amounts of driving torque. The driving and braking forces are responsible for large amounts of weight transfer and squat and dive actions.

Lateral friction – when travelling in a straight line with no lateral, or side, forces on the vehicle the tyres have a true rolling motion over the road. When cornering centrifugal force exerts a lateral force on the tyres, this in turn gives rise to slip angles. That is the difference between the line of the wheel and the actual line of travel. Side winds and variation in the steering angles can also generate slip angles. If both front and rear slip angles are equal then the vehicle has neutral steer characteristics. If the slip angles are greater at the front than at the rear then the vehicle will under steer. If the slip angle is larger at the rear than the front, then it will over steer. Under steer is a stable condition and therefore desirable in a road car, but not necessarily a race car. Variables which bring about changes to the lateral friction are:

- Load distribution, a heavy load at the rear will increase the rear wheel slip angles and cause over steer.
- Tyre aspect ratio (height/width), tyres with an aspect ratio < 0.70 allow high cornering speeds.
- Road conditions, typical co-efficient of friction for a good road are 0.80 to 0.85.
- Changes to the track width brought about by suspension movement can change the slip angles.
- Changes in the load distribution, such as brought about by heavy breaking, alter the vertical forces on the tyre, this alters the amount of lateral friction and hence the vehicle's stability.
- Camber changes, both wheel and road camber, alters the lateral friction.

Load-carrying capacity – before calculating the load-carrying capacity it is important to define kerb weight (KW) and gross vehicle weight (GVW). Some literature uses the more technically correct term vehicle mass, for all intents and purposes the two terms are synonymous, the use of the term weight is unlikely to go out of colloquial use in this country or America. By definition of EU and DIN standards the kerb weight includes: the chassis; body and trim; engine; gearbox; all ancillary items such as starter, alternator, battery, fuel system, exhaust system and cooling system; optional items such as sunroof and air conditioning; all the oils and fluids including 90 per cent of the maximum fuel; all the mandatory safety items, namely, jack and brace, spare wheel, warning triangle, first aid kit and spare bulbs.

EU type approval regulations, which do not currently apply to one-off cars, require that the maximum mass of drivable vehicle (MDRV) must be stated, this includes an allowance of 68 kg for the driver and 7 kg for luggage.

The load-carrying capacity, or payload, is the difference between the permissible gross vehicle weight, as determined by the vehicle design, and the kerb weight.

Roll – is the difference between the overturning moment and the righting moment. The speed at which a vehicle will overturn depends upon the height of the centre of gravity, the track width and the radius of the corner. The roll centre of the vehicle may be above or below the road surface; this depends on the layout of the suspension mountings and the angle of the suspension arms.

Sprung and unsprung mass – the sprung mass (SM) is that mass above the springing medium, the unsprung mass (USM) is that below. Part of the mass of the suspension components may form unsprung mass, the remainder sprung mass. It is desirable to keep the unsprung mass as low as possible in relation to the sprung mass so that the amplitude of the force generated when the wheel hits a bump is kept to a minimum and cannot therefore cause driver or passenger discomfort. Three concepts related to the sprung to unsprung mass ratio which affect the ride quality are:

- **Wheel hop** – this is usually a resonant condition of the unsprung mass which is at a different frequency to the frequency of the body/chassis. On classic cars this can be felt through the steering. Makers of the Vincent, a replica of the classic 1930s MG, fit FX Taxi wheels to an otherwise lightweight suspension system to generate an amount of wheel hop, sufficient in their words to give a vintage feel to the suspension. The wheel hop must not be such that it adversely affects the handling of the vehicle. With a live rear axle, tramp can develop when cornering at speeds exceeding the design figure, this feels like wheel hop, indeed it is a sideways form of wheel hop generated by the driving forces.
- **Wheel patter** – the low amplitude wheel hop, particularly noticeable on washboard type surfaces, such as concrete sections of motorway.
- **Boulevard jerk** – is caused by the **stiction** in the suspension system, this gives a jerky ride on a fairly smooth surface. On a bumpy road this cannot be felt because of the suspension action in dealing with the bumps.

Track width – (TW) is the distance between the centre of the tyre tread contact patches on each axle. It is common for there to be a small difference between the front and rear track widths. The track width has a major influence on the vehicle's cornering behaviour and the amount of body roll. The track width should be as large as possible within the constraints of the body/chassis width. The **overall width** (OW) is the width of the outside bodywork of the car at its widest point. The track width ratio (k_2) is an indication of how well the suspension fits the body chassis and is calculated by: $k_2 = TW/OW$.

The Motor Vehicle Construction and Use Regulations (C&UR) require that the body work completely cover the wheels and tyres, therefore it is impossible for the k_2 figure to be equal to (=) or greater than (>) unity. The wheels and tyres need running clearance to cope with the build-up of mud or snow and allow for minor variations in tyre sizes when replacements are needed. In bump and rebound conditions the TW may vary, so there must be body/chassis clearance to allow for this.

On many independent suspension systems the bump and rebound movement of the suspension causes changes in the TW. This is because of the geometry of the suspension pivots. The problem is that variation in track width causes lateral forces which give rise to variations in slip angle (α). This leads to an increased rolling resistance and adverse affects on the vehicle's directional stability which may include steering effects. That is **bump steer**. The loss in performance and the detrimental effects on the steering at high speed are to be avoided. Changes in track width also cause tyre noise and increases tyre wear. The tyres used on motorsport vehicles are expensive, so excessive wear generating geometry should be avoided, and for environmental reasons vehicle noise should be kept to the minimum.

Wheel alignment – this is the relative position of the road wheels. The wheels are usually set with a predetermined value of static toe-in or toe-out so that under dynamic cruising conditions (30 to 70 mph or 50 to 110 k/h) the wheels will become parallel to each other and so have no excessive wear on any one edge. Conventional layout vehicles, that is ones with a front mounted engine driving through a gearbox and propeller shaft to a rear axle, usually have a small amount of toe-in, typically 2 or 3 mm measured at the wheel rim. The positive wheel camber tends to make the front wheel want to peel outwards at cruising speed, so that the static toe-in is changed to a parallel position. With radial ply tyres the toe-in can be set to a minimum value, the final recommended setting would depend on tyre-wear observations.

To ensure that the wheels have true rolling motion on corners the Ackermann principle is usually designed into the steering arm layout. That is so that the inner wheel turns through a greater angle than the outer; typical figures are that the inner wheel turns through 22

FIGURE 2.3
Turntables for measuring toe-out on turns

degrees when the outer is turned through 20 degrees. This difference is referred to as toe-out on turns.

On vehicles with independent rear suspension (IRS) a small amount of toe-in is given to the rear wheels to allow for suspension movement caused by positive rear wheel camber.

FIGURE 2.4
Close up of scale on turntable

Wheel base (WB) – is measured from the centre of the front axle (imaginary) to the centre of the rear axle. The overall length (OL) is measured from front bumper to rear bumper. Both the wheelbase and the ratio of the wheelbase to the overall length (k_1), k_1 = WB/OL, are important variables in suspension design. A long wheelbase relative to the overall length of the vehicle allows for the accommodation of passengers between the axles, so the floor can be flat in the foot well and the seat cushion height has less special constraints. It reduces the affect of load positioning in the vehicle; this includes the position of the engine and gearbox. It reduces the tendency to pitching, especially on undulating country roads which in turn allows the use of softer springs which tend to give passengers a more comfortable ride. With the reduced overhang there is less **polar inertia**, which improves the **swerveability** of the vehicle. The length of the wheelbase affects the turning circle for any given input of steering angle. American car companies have set ratios for wheel base and overall length; this is to give a particular aesthetic to their full size cars – for example the Lincoln. On the compacts and sub-compacts (ordinary European style cars) this does not apply.

TRY THIS

Drive two different vehicles and compare the feel of them on the road using the terms you have learnt about.

SUSPENSION TYPES

FIGURE 2.5
Wishbone lower link

Beam axle

The beam axle type of suspension (also called rigid axle) is the simplest and the oldest type, it is frequently supported using semi-eliptic leaf springs. Beam axles are often used at the rear of front engined race cars in the form of a live-axle, that is one which transmits the power through half-shafts inside. Beam axle front suspension was used on cars of the 1930s and 1940s but it is not a viable option for front suspension on a race car because of the effect that hitting a bump has on the steering at high speeds. Racers used to cut the front beam axles in half and convert them to swing axles by the addition of an central pivot point. Interestingly, this was used by Ferrari as well as on a great number of Ford E93 based kit cars. The problem with beam axles is that if one wheel hits a bump both wheels are tilted, so the castor, camber and king pin inclination are all altered. A live rear axle is a possible choice for the rear suspension of a race car as it allows the use of a variety of engines and gearboxes with a conventional layout chassis, also sufficiently strong rear axles are available for transmitting the 300 to 400 bhp which is needed to get the quarter mile well below 10 seconds.

Independent suspension

The beam axle is non-independent, most other suspension types, where each wheel can move separately, are referred to as independent suspension.

MacPherson strut

A MacPherson strut suspension is a concentric arrangement of coil spring and damper unit which is also the suspension upright member. It is used on many cars for both front and rear suspension. It allows for a large amount of suspension travel and can therefore provide a soft ride with a long stroke damper action. The mounting parts are well apart and so the load can be spread across the body/ chassis components. This allows plenty of space between the struts for either an engine or luggage space, on the negative side a considerable height is needed to accommodate the struts, this constrains the vehicle height to a minimum figure. MacPherson strut suspension, which by its nature needs to be mounted with a small angle of inclination to the vertical, tends to give a large amount of track width variation on bump or rebound, this creates a lateral force which is taken by the plunger arm such that it is susceptible to wear and may develop a coarseness in the steering. An anti-dive action is not usually incorporated.

Vertical pillar strut

The vertical pillar strut suspension is initially similar in both action and appearance to the MacPherson strut design, but the sliding member is such that the action is truly vertical so that the tyre maintains a constant contact patch and track dimension. It is also designed so that the damping, suspension springing and track control actions are separate. This suspension is usually only to be found on Morgan sports cars.

Wishbone

The usual arrangement is a double wishbone, the upper wishbone being shorter in length than the lower one. Anti-dive is achieved by inclination of the arm axis. This design requires separate dampers and springs. The position of the mounting point is critical in terms of vehicle roll and other suspension and steering actions. The space between the inner pivots of the wishbones can constrain the engine bay space; the height is such that it is usually no higher than the road wheels.

Multilink

This is an adapted version of the MacPherson strut which controls the wheel movement to give a constant track and a more robust mounting arrangement. It is a feature of many Japanese or other far-Eastern saloons.

Trailing arm

Trailing arm suspension can be used on both the front and the rear, but it is normally only used on the rear. The problem is that the arm tends to lift under hard acceleration and traction is lost. Vehicles which drive through rear trailing arm set ups have very poor grip on snow and ice. Trailing arms are now used only on dead (no drive) rear axles. They ensure that there is no change to neither the track width nor the camber angle.

TRY THIS

Make a list of motorsport vehicles and identify the types of front and rear suspension used – which is the most popular?

SUSPENSION MEDIUMS

A number of different mediums, or materials, can be used to form the suspension. The displacement of the medium is the action of the suspension. Of importance is the amount of displacement and the rate of displacement, this depends on the elasticity of the medium and a number of other factors related to the design of the suspension. On a rally car it is necessary to have a moderately large amount of displacement in the form of available suspension travel because of the size of the components and to give reasonable ride height and

FIGURE 2.6
Tubular links

comfort over a variety of road surfaces. On a circuit car the suspension movement may be as little as 2 mm. The rate of displacement, measured in terms of suspension rate, the load necessary to compress the medium by a unit distance, in race car terms this is usually referred to in pounds per inch (lbf/inch).

Hydrolastic

The Hydrolastic (registered trade name) fluid is used to transmit the forces from the suspension arm to the thick rubber diaphragm which actually does the job of springing. The shock absorber function is built into the Hydrolastic displacer unit. The usual arrangement is to connect the displacers front to rear on each side. This reduces pitching and the bump effect when one wheel hits a road irregularity. Hydrolastic suspension could be used on a race car, but it is generally restricted to light weight (class A) saloons.

Hydropneumatic

The suspension medium is gas which is compressed inside a spherical chamber by means of a diaphragm against which the oil-based fluid is pressed. The fluid transmits the force between the suspension arm and the diaphragm. The hydropneumatic suspension may be fitted independently to each wheel; or on light vehicles the suspension is often connected between the two sides on each axle. That is, they are connected transversely. On more sophisticated cars, like the Citroen, all the suspension units are connected through a valve system. The fluid is pressurised by a pump so that the suspension can be raised and lowered and constant body height may be kept when the vehicle is fully laden. On pumped systems the length of suspension travel can be quite long.

Torsion bar

The torsion bar is favoured for its light weight; but it needs a long installation length. The length factor is to some extent compensated for by the minimum amount of height which is needed. Torsion bars generally have only small amounts of suspension travel available; they are not, therefore, suited to custom cars. They can be found on small Fiats and the now classic Morris Minor. The mass of the torsion bar is entirely sprung weight as it is mounted at each end to the body/chassis.

Rubber cone

Used solely on the old Mini as a main spring, and occasionally on other vehicles as a secondary spring. In its secondary spring role it is favoured by caravan towers to prevent the rear suspension from bottoming and provide a variable rate spring for a fully laden vehicle. The load carrying capacity is limited to a small figure and the suspension travel distance is very small.

Coil spring

Helical coil springs can be made to allow for a long length of suspension travel. They can be made in a range of spring rates including variable rates. Coil springs can be fitted to suspension arms in a large variety of ways including concentricity with the shock absorbers.

Leaf spring

Leaf springs can also take on the role of suspension member as well as that of spring medium. They may have one leaf or a combination of different length leaves laminated together to form the spring. The leaf springs are usually made from medium carbon steel, but plastics and composites are optional materials. Leaf springs may be used in either semi-elliptic or quarter-elliptic forms. Leaf springs are easy to attach to both the axle and the chassis; they also may provide a medium range length of travel.

Ladder bars

A ladder-shaped linkage used between the chassis and the live rear axle. It operates in a similar way to a trailing arm. This set up is often used on dragsters as the length can easily be changed and mounting to the axle is simple.

SUSPENSION COMPONENTS

Four bar

Four separate linkage bars between the chassis and the live rear axle. They operate in a similar way to a trailing arm, but the mounting points are spaced apart to transmit the driving and braking forces more evenly. They also offer a high resistance to the axle twisting under heavy acceleration.

Watt linkage

A centrally pivoted linkage system which locates the wheels transversely; this is used with IRS on some vehicles.

Panhard rod

A single transversely mounted rod used on both live and dead rear axles to absorb lateral forces and reduce the risk of axle tramp.

STEERING GEOMETRY

Non-steered wheels – on a live, or a dead, axle the wheels will both be perpendicular and with no toe-in, nor toe-out.

On non-steered independent rear suspension (IRS) a certain amount of camber (positive or negative) may be incorporated to maintain maximum tyre contact on cornering – especially short circuit cars. Toe-in or toe-out will therefore be needed to prevent unnecessary tyre scrub.

FIGURE 2.7
Race car set-up for corner weight checking

Camber – the inclination of the road wheel when viewed from the front of the vehicles. Leaning out at the top is positive camber; leaning in at the top is negative camber. The camber angle can be 0 to 8 degrees. Road cars usually run small amounts of positive camber; single seat race cars are likely to have large values of negative camber – the reasons for camber are to accommodate camber changes on bump steer and maintain maximum tyre contact patch size.

Castor – this is the rearwards inclination of the suspension upright, this gives the self centring action of the vehicle when you come out of a corner.

SAI (KPI) – the inclination of the swivel axle (better known by the old term king pin, usually the suspension upright on a race car). This usually inclines inwards at the top end by 0 to 4 degrees.

Toe-in and **toe-out** – the static setting of the road wheels looking from the top. Toe-in means that they are closer at the front; toe-out is that they are wider apart at the rear. Cars that drive through the rear wheels usually have toe-in; FWD cars have toe-out. The reason being that under steady state cruising the driving forces will move the wheels into a neutral position to ensure no tyre scrub. A typical figure for toe-in or toe-out is 0 to 3 mm (0 to 1/8 inch).

Ackermann principle – this is used to reduce tyre scrub on turns, giving as near as possible true rolling motion on turns. This is achieved by lining up the centre of the track rod end (TRE) on an imaginary line between the suspension up right pivot centre and the centre of the rear axle. So that as the outer wheel turns, the inner wheel turns through a bigger angle. Typically when the outer wheel turns through 20 degrees the inner wheel will turn through 22 degrees.

Toe-out on turns – this is the correct test for Ackermann:

- Place the vehicle with the front wheels on turntables and the rear wheels on thin chocks to ensure that it is level.
- Set both wheels to straight ahead position – check the steering wheel.
- Set the turntable scales to zero.
- Turn the steering to the left until the outer wheel, the one on the right, has turned through 20 degrees.
- Check the angle of the inner wheel, the one on the left, it should be 22 degrees or as recommended by the manufacturer.

If the Ackermann angle is incorrect the vehicle will scrub off tyre tread in a similar fashion to incorrect wheel alignment. Ackermann faults can be caused by: bent steering arm, out of centre steering rack fitting, steering rack adjusted to one side. However, be careful and check the manufacturer's data, not all vehicles use Ackermann – they just scrub off tyres.

TRY THIS

Carry out a steering and suspension check on a vehicle.

Slip angles – this is the difference between the direction of travel of the vehicle and the angle of the wheel.

Self-aligning torque – the torque exerted by the castor angle which tends to pull the vehicle into a straight line. On circuit cars the castor angle is kept small, typically 0 to 3 degrees to keep the steering light. Heavy luxury cars may have up to 8 degrees of castor to keep them on a straight line at speed on the motorway.

Over steer, under steer and neutral steer – are the three characteristic steering tendencies of vehicles. Neutral steer is the situation when you go into a corner having applied an amount of lock and the vehicle follows the planned path with no more steering changes. Under steer is when you enter a corner then have to apply more lock to negotiate it completely, in other word the vehicle does not follow the front wheels. Over steer is when you go into a corner and then have to remove lock and the vehicle feels like it will roll over. Tyre types, sizes and pressures play a big part in this area of handling. Also, changing the load in the vehicle will alter the handling characteristics – a vehicle with four passengers will be much more difficult to drive than one with just two occupants.

FIGURE 2.8
Corner weight pad

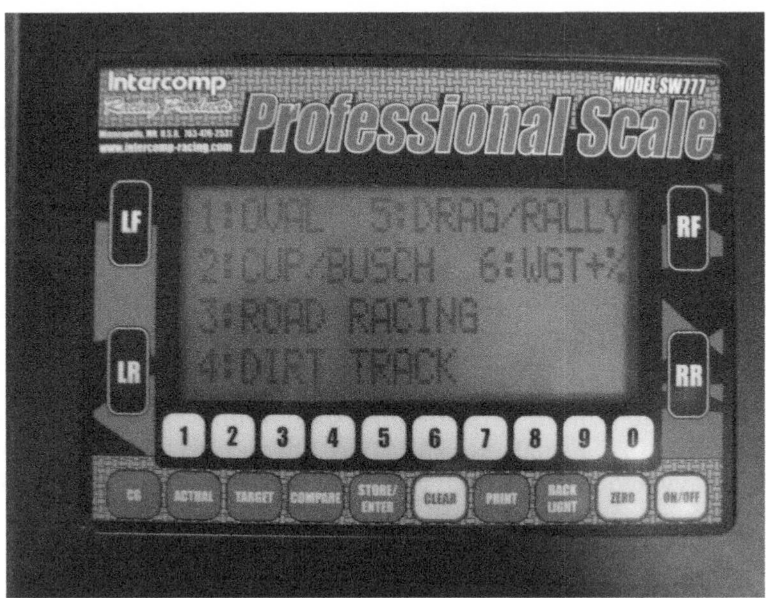

FIGURE 2.9
Read out screen for corner weight gauge

Rear and four-wheel steer – some vehicles have been made with rear wheels which move a small amount when the front wheels are steered.

Hydraulic power steering – relies on hydraulic fluid pressure to move the steering rack, or in some cases a steering box. Some systems are speed sensitive, that is the amount of power steering assistance decreases with the road speed. In its simplest form this actually just measures engine speed, the argument being that engine speed correlates to road speed. The more complex systems pick up road speed information from the ECU.

The diagnostic procedure to be followed is given in Table 2.1.

TABLE 2.1 Hydraulic power steering fault finding

No.	Test	Mfg pressure	Test pressure	Comment
1	Check for fluid leaks	n/a	n/a	only use specified fluid
2	Check for noises	n/a	n/a	check for bearing wear
3	1,000 rpm full left	45 psi (3 bar)		typical figures
4	3,000 rpm full left	75 psi (5 bar)		
5	1,000 rpm full right	45 psi (3 bar)		
6	3,000 rpm full right	75 psi (5 bar)		

FIGURE 2.10
Coil spring rate testing – hydraulic pressure

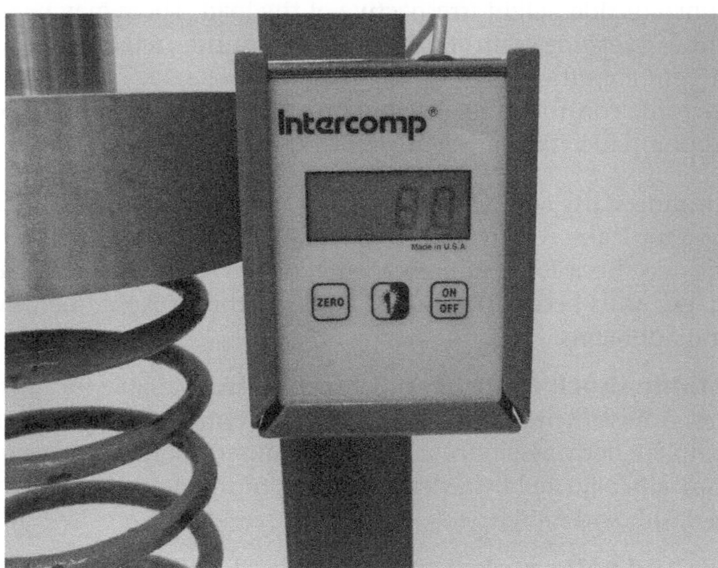

FIGURE 2.11
Digital read out for spring rate force

RACER NOTE

Driving fault free cars helps you identify faults on others.

Electronic power steering – is controlled through an ECU which monitors:

- vehicle speed
- force applied by the driver
- steering angle position
- rate of acceleration of steering angle – how fast the steering wheel is being changed.

Fault finding is usually limited to using a fault code reader.

SUSPENSION ADJUSTMENT

Adjustable ride height – there are a number of different systems, namely:

- Self energising – an independent unit, like a bicycle pump, is mounted between the axle and the body/chassis. As the suspension moves it pumps up internal pressure. A set of internal valves control the pressure to match the external load and the relative ride height.
- Pneumatic suspension with height control – this has a height control sensor which measures the effective ride height and allows more fluid to be pumped into the suspension units to effectively raise the suspension. Under normal conditions this will maintain a constant ride height irrespective of the load. There may be some form of over-ride control to allow different ride height for different conditions.
- Electronic controlled suspension or active suspension, the valve and pump mechanisms being controlled by an ECU.

Fault finding starts with using a fault code reader if appropriate, then checking for leaks, which are the usual failure. Some suspension systems need occasional re-pressurising, or the replacement of the diaphragm units because of internal leaks or the weakening of the rubber diaphragms.

Adjustable shock absorbers (dampers) – are available for most vehicles. Adjustment may be by a small screw at the base or twisting the complete body assembly with a 'C' spanner. Some dampers are electronically adjusted by a stepper motor, or solenoid, controlled from a dashboard switch.

Rocker and bell crank systems – are used on most single seaters to reduce unsprung weight and put the damper into a cooling air flow. It is easy to check them visually.

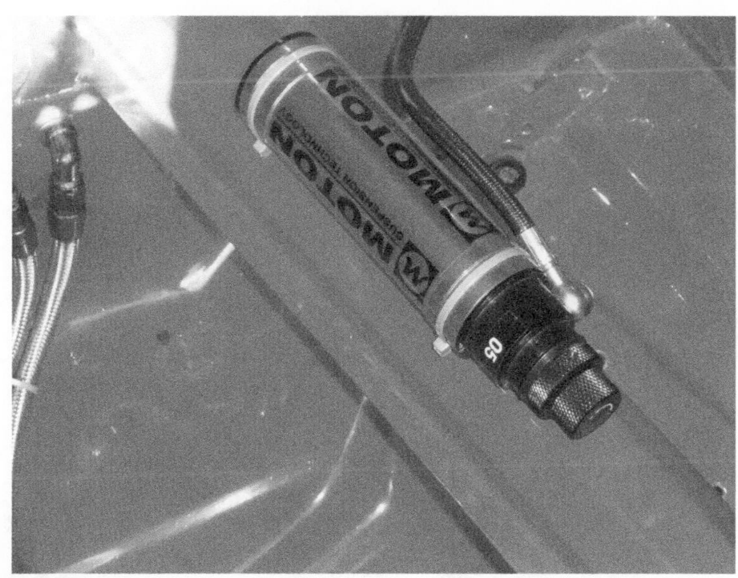

FIGURE 2.12
Damper (shock absorber) reservoir

Push-rod and pull-rod systems – in a similar way to the bell crank system they allow the reduction of unsprung weight and allow remote positioning of components.

Adjustable axle location – the use of rose joints and threaded tie-rods allows suspension components to be adjusted to give exact set-up steering geometry.

TRY THIS

Drive a vehicle with adjustable suspension and make some adjustments – how does its handling change – do this on a track only.

LEGAL REQUIREMENTS

Road safety and track safety – in the wrong hands any vehicle is a lethal weapon. By its very nature a motorsport vehicle is very dangerous, that is because of its high power-to-weight ratio and its twitchy handling and unforgiving controls. To put safety into perspective consider the following facts based on typical figures for the UK:

- 70% of road accidents are by drivers aged under 24 years old
- 12,000 people are sent to jail each year for driving offences, 2,400 of these are for driving under the influence of drink or drugs
- six people die each day on the roads, hundreds are injured.

FIGURE 2.13
Adjustable strengthening link on front suspension

In the USA and other countries the figures are proportionally similar, as a general guide in the UK there are about 30 million vehicles, in the USA the number is about 250 million.

The risk is not just from you as a driver; but includes other drivers colliding with your car and the risk of mechanical failure causing an accident. Think safe, the following positive actions will help to keep you alive:

- Check the vehicle over before you race, or when you take over a vehicle which you are not familiar with, this may include: a spanner check, oil/coolant/fluid level check, wheels and tyres, door/boot/bonnet catches, and general check for condition.
- Carry out a dashboard check and start up procedure for the vehicle – remember some engines need to be started in special ways.
- Wear appropriate race/rally clothing – Nomex underwear, driver's overalls, driving boots, helmet, visor and gloves.
- Fasten the seat belt correctly.
- Follow all flag and light instructions.
- Watch out for and avoid other drivers not driving safely.

Lighting – on the road it must comply with the appropriate regulations. If a rally event goes into another country then it must comply with those regulations too.

HISTORIC RACER NOTE

In 1965 the Monte Carlo winning Mini Cooper S was disqualified for having the wrong headlamps.

Tyres – if used on the road these must be road legal, you can not use slicks on the road. For race events the tyres must comply with the race regulations. It is normal procedure for Formula type events to have control tyres – that is one particular make, size and tread pattern. There may be two choices, namely wet weather tyres and dry weather tyres. The tyres must be fitted and inflated to meet strict guide lines. It is normal to do this with the wheels and tyres separately before scrutineering when they will be fitted with a seal. As tyres run best when warm the use of tyre warmers is a good idea. Be aware that tyre temperatures will be very different, and therefore so will handling and lap times, in different weather conditions – February testing at Cartagena will be much faster than at Snetterton.

RACER NOTE

Recently, one Formula 1 team had a choice of 1,500 different tyres.

Steering – must be free from play, and not damaged.

Brakes – must comply with regulations and be similar to others in same class. It is important that braking power is even for each axle set, and that front rear balance gives straight line braking in a straight line under dry conditions. Pads and shoes must have sufficient friction material for the event.

Trials cars have fiddle brakes – that is there is a separate lever for each rear wheel; some motor cycles have inter-connected front and rear brakes.

Seat belts – for competition use it is essential to have multi-point seat belts, the current trend is for six point mounting, although you will also find four and five point melting seat belts.

Emissions – there is a lot of talk and publicity about emissions. To put it into perspective, the amount of pollution generated by motorsport in the UK equates to that generated by one jumbo jet flying from the UK to Australia, and the aeroplane puts the pollution in the most dangerous place. However, we must ensure that we do not cause unnecessary pollution from our exhaust, or elsewhere. Where possible use catalytic converters, and for diesel engines use after-burners. Also, keep the noise down as much as possible. Most events have a scrutineering area for testing exhaust noise and many can test exhaust emissions. Thruxton Circuit has *church hour* on a Sunday morning; no racing or testing is allowed for this hour.

Rally cars may be fitted with two engine ECUs; one giving an over rich mixture to keep the turbocharger working for maximum acceleration on special stages, the other to give road legal emissions when driving between stages.

HISTORIC RACER NOTE

Depending on the vehicle and its age, the regulations may vary, for example:

- Steering may have a limited amount of play in the straight ahead position.
- Brakes may have low (by comparison to current vehicle) efficiency.
- Seat belts may not be needed.
- Emissions limited to no visible smoke from exhaust.

EuroNCAP – the European New Car Assessment Programme is an independent crash testing organisation standard. It is used by all the major car manufacturers in Europe. The Australian car makers are now adopting it as are the car makers in the USA. Currently the USA uses the National Highway Traffic Safety Administration (NHTSA) standards. It is not a legal requirement to comply with any of these standards; but they are an excellent measure of how the vehicle will perform when involved in a collision. Based on their performance the vehicles are given a *star rating*. Five stars is the highest rating. Formula car tubs are sometimes tested in the same way. There are four major tests: Front Impact Test, Side Impact Test, Pole Test and Pedestrian Test.

MOT – in the UK this refers to the MOT Test Certificate, its name came from the Ministry of Transport, now Vehicle Operating Standards Agency (VOSA). However the name MOT still appears on the certificates in deference to the popular demand.

The MOT is a 30 minute examination which is now done on-line so that all MOT certificates are kept on a database allowing re-taxing of the vehicle over the telephone. Vehicle insurance is also kept on a database so that VOSA know which vehicles have tax discs, insurance and MOTs.

A similar situation exists in the USA and other western hemisphere countries. Less developed countries still require tax and insurance, but the testing and databases are not as well developed as administration tends to be at a more local level.

Competition requirements – general competition requirements are covered in the **MSA *Bluebook***. Specific regulations (**regs**) for each class, or type of racing, are given by the organisers; usually these are published on the internet to reduce the use of paper and postage costs.

FIA – Federation International Automobile, the world governing body for motorsport. However it is more important in Europe and the emerging motorsport countries like the UAE and Japan than the USA. Many form of motorsport do not recognise the FIA, preferring to have independence. Sometimes this leads to arguments and conflicts.

MSA – Motor Sport Association, the leading motorsport authority in the UK, they are a member of the FIA. The local motor clubs form the membership of the MSA along with licence holders. The MSA is responsible for most of the racing in the UK. The MSA issues racing licences to both individuals and race circuits. It organises some premier events like the British Grand Prix. Motor cyclists have the **Auto Cycle Union** (ACU). Kart racing is administered by the MSA; but drag racing is not, it is done through the **National Hot Rod Racing Association** (NHRA). Probably the oldest club is the **Motor Cycle Club** (MCC) which may be confusing as it organises off-road car trials as well as motor cycle trials. The USA has a number of clubs the best known probably being **National Association for Stock Car Auto Racing** (NASCAR) which operates in the UK too.

Bluebook – published annually by the MSA which sets out the rules and regulations of racing in the UK.

ARDS – the full name, which is rarely used, is the Association of Racing Driver Schools. ARDS administers and certificates the qualification for a racing licence. The normal procedure is that before starting racing you have a medical examination, then you spend a day and the cost of a couple of tyres on a one day training course with an ARDS registered racing driver school. At the end of the day you take a short written examination – multiple choice questions and a number of observed laps around the circuit. The purpose of the ARDS is to ensure that you will be safe on the track and observe the flags and appropriate directions. If you pass your medical and ARDS you can apply to the MSA for a competition licence.

Competition licence – issued by the MSA following your medical and passing your ARDS. There are a number of different competition licences for different classes of racing. For non MSA events, like some drag racing, you can join the NHRA, or in some cases pay on the day for a temporary licence. The difference is that in MSA events there are usually many competitors on the track at the same time; drag races are run against the clock, the risks are different.

Scrutineering – the checking of the vehicle before an event. The scrutineer checks to see if the vehicle complies with the regulations relating to the event and this includes general safety too.

Driving on the road – to drive a vehicle on the road, both the vehicle and the driver must comply with all the regulations. For example:

- safe legal vehicle
- MOT if needed
- insurance

- driver's licence
- road tax.

Driver's licence – to drive a vehicle on the road in the **UK** you must be over **17** years old. You can apply for your licence and your test up to three months before you are 17. In the **USA** the age varies in each state, typically **14** in the mid-West and **18** in New Jersey. There are also rules in the USA that require you to have a learner's permit before a full licence will be issued.

Road tax – is essential if the vehicle is being used on the road. Road Tax can be bought at the Post Office, on-line, or over the telephone. To apply on-line, or over the telephone, you must have **insurance** registered on the insurance database and if the vehicle is over three years old, a new style **MOT certificate**, these are also registered on the VOSA database. VOSA will post out the tax disc within about 4 days.

Insurance – legally required for a vehicle to be used on the road in any country, it usually does not cover any racing or rallying. Also motor trade insurance does not cover competition vehicles as standard. Special insurance is available for racing and rallying, there are different levels of cover, and these may be very specific, typical examples are:

- transporting to an event
- testing
- at event (not on track, not racing)
- on-track (not racing)
- on-event (racing).

Check out the fine detail of any insurance before buying, or going to an event.

Company regulations (smoking/alcohol) – it is illegal to smoke in a company-owned vehicle in the UK, even if it is your own company. It is also illegal to smoke on any business premises. Generally speaking smoking, the use of alcohol, or drugs is banned in all motorsport premises and vehicles in the UK, the USA and most western hemisphere countries.

Highway Code – you have a legal requirement to fully understand and abide by the Highway Code in the UK. In the USA the highway regulations do vary in detail between states for things like (traffic) stop signals and lane usage; but similar principles apply – that is taking care, signalling and obeying the signage. European countries tend to follow a similar pattern.

Vehicle care – as a technician there is no need to mention that you have a duty of care of any vehicle in your charge.

HEALTH, SAFETY AND THE ENVIRONMENT REQUIREMENTS

Remember what you learnt at Level 2.

Motor racing can be dangerous, so no risks off the track – remember to document all your actions.

THINK SAFE

Both **employee** and **employer** are responsible for third parties. **Vicarious liability** means that the employer is responsible for the actions of the employee as well as the employee being responsible for any action – this means that all parties may be liable for prosecution.

COSHH – Control of Substances Hazardous to Health – you should have a file for each of the substances that you use in your workshop showing the hazards and how to deal with them – your suppliers will be able to provide this data; and most companies have it available on their web site.

Risk assessment – assessing each action before doing it.

Disposal of waste materials – is simple if you follow a few basic rules – to be environmentally conscious is very important, doing this correctly will save costs in the long run and give a good image to motorsport in general.

THINK SAFE

Disposing of motorsport equipment wrongly can cost you in fines and you must also be careful not to give away clues as to your racing secrets.

Never put in a land fill bin anything which can be re-cycled, in many cases it is against the law, it will depend on your local authority, and it costs nothing to ask for advice; but ensure that you follow it. Examples typical of most areas in the UK and USA are shown in Table 2.2.

THINK SAFE

Professionally and personally over the past three years I reduced my land fill waste to a quarter of what it was. As well as recycling everything I can, I also challenge the use of unnecessary packaging of items, that is asking why things are packed in a certain way and requesting changes were I can. I also look for the best choice of alternative materials being specified and re-use of materials were appropriate.

TABLE 2.2

No.	Item	Where to dispose	What happens to it	Comment
1	old units	return to manufacturers	may be overhauled, re-manufactured	
2	electrical components	specialist disposal at ERS	precious metals removed for re-use	
3	tyres	specialist disposal through supplier	shredded for use in road surfaces and safety surfaces	specialist companies will collect these
4	batteries	specialist disposal at ERS	precious metals removed for re-use	specialist companies will collect these
5	oil and fluids	specialist disposal at ERS	converted in to paint stripper or burnt to generate heat	specialist companies will collect these
6	paper and card	collected by local authority	re-cycled into paper again	
7	wood	specialist disposal at ERS	burnt in power stations to make electricity	
8	metal items	weigh in for scrap	melted down for re-use	
9	glass	specialist disposal at ERS	melted down to make more bottles	

Note: (ERS) Environmental re-cycling site

THINK SAFE

Ensure that the workshop, the transporter and all the vehicles are equipped with the correct fire extinguishers and that they are live and tested before every event.

TECHNICAL INFORMATION

Information sources – technical information for road vehicles is not always easy to find; for race vehicles it can be impossible. As a race engineer you should keep your own records and notes as well as those needed by the customer, team, or other service provider. The use of a digital camera is recommended as the information can be uploaded on to your computer. If you are purchasing a mobile phone with a camera, then look for one with zoom and flash built in. The macro facility too is useful for photographing small parts and pages of data. Sources of information may include:

- manufacturer's workshop manual
- independent type workshop manual (such as Haynes)
- driver's handbook
- CD based manual
- parts manual
- on-line data source (such as Autodata)
- information from test bench manufacturers (such as Snap On)
- company records
- data acquisition system
- vehicle log.

Maintenance of records – work on any vehicle must be documented. You must maintain an audit trail to ensure that you have both a technical and a legal record of what you have done. Depending on the type and level of racing, you are likely to keep records in one, or more, of the following ways:

- vehicle service record
- company job card
- company file card
- Electronic Customer Record Management System (CRM) also called Customer Relationship Management
- data log – hard or soft
- vehicle log – hard or soft.

Legal requirements relating to data – you must keep detailed records of all your work for the following purposes:

- ensure that the vehicle complies with the relevant competition regulations
- meet the needs of the driver and/or vehicle owner
- satisfy the requirements of HM Revenue and Customs (UK) or other tax and VAT requirements in other countries
- meet HASAWA criteria – in case of an accident
- ensure compliance with the Data Protection Act (UK and most other countries).

The law requires that you keep this data safe and do not disclose it to third parties. Written (hard copy) records must be kept in a locked file – usually these are fire resistant too. Electronic (soft) data must be password protected and only be accessible to authorised parties. The Data Protection Act, which is universal in most countries, makes these stipulations which are covered by criminal law as well as civil law.

FAULT DIAGNOSIS

This topic is largely covered in Chapter 4 'Inspection'. However, specific to the vehicle chassis the following procedures ought to be borne in mind when trying to diagnose faults:

1. Where possible, and appropriate, take readings using the fault code reader to interrogate the chassis ECU. This is always the first stop if possible.
2. Following an off, no matter how light, check the steering geometry, the chassis alignment, and corner weights. If the car was not damaged by the off, it could have been a faulty steering, suspension or chassis component that caused the off in the first place.

Although for most purposes checking the toe-in/toe-out with optical gauges is good, and a manual check of castor, camber and SAI with a bolt-on gauge will suffice. The use of a complete 4-wheel system will show up more faults – that is relative alignment of all four wheels. The corner weights can also indicate incorrect suspension adjustment, or faults.

If it is suspected that the chassis is bent then carry out either a jig test of drop test with plumb-bob and chalk.

SET-UP PROCEDURES

Pit garage – using a pit garage will usually have an on cost, whether it is a track day, test day or race day. Costs vary with time of year and circuit. The advantage of a pit garage is that you will have a flat floor, shelter, be right next to the track and can usually set it out and maintain security. Depending on the size of the vehicle you may be able to accommodate more than one car. Most pit garages have transporter, or trailer, access from the non-track side.

Paper based systems – have their place in *clubman* and *historic* vehicle events where the changes to the vehicle are likely to be small and the set-up follows pre-set patterns or methods. Then it is a matter of recording only limited variables like tyre pressure and fuel consumption.

Laptop systems – most competitive teams and individuals use a laptop computer based system. This allows lots of data to be recorded quickly and retrieved for analysis and comparison.

The *Intercomp Race Car Management* software has six heading tabs, these are:

Geometry – details of the suspension including: castor, camber, SAI, which components and vehicle height
Setup – details of spring and damper settings
Weights – corner weight information showing CG
Track – details of track and weather conditions
Race – race times/speeds, section times and comparative information
Tyres – compound, pressure and temperature.

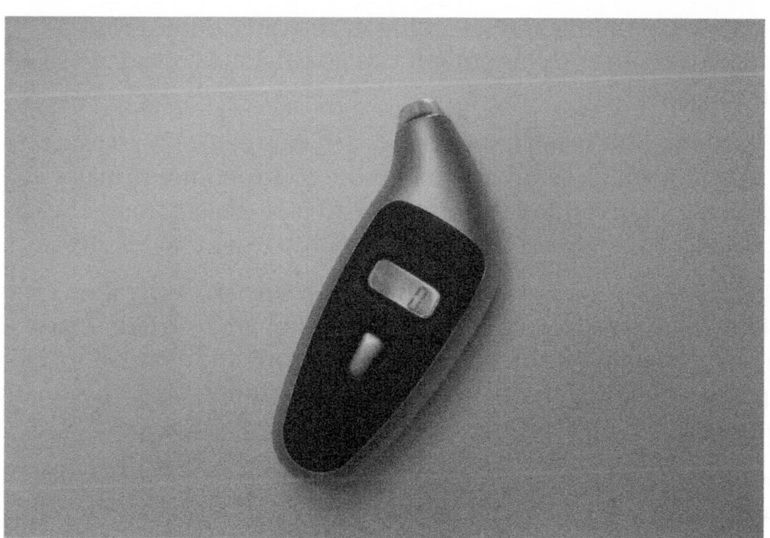

FIGURE 2.14
Digital tyre pressure gauge

Of course you can record this information on paper manually; but the low cost of software and easy access to laptops, coupled with the speed and accuracy of such a system makes it ideal.

Data logging – consists of three main areas:

- The set-up data of the vehicle and how this changes during practice and racing.
- The performance of the vehicle under practice and race conditions, this will be logged against individual drivers.
- The behaviour of the vehicle on the circuit.

These areas may be separate, or they may be integrated.

Data analysis – once you have got the data you have to analyse it. That is, you need to break it down into small manageable chunks and try to make sense of what is going on. The big problem at the beginning of any season, or on delivery of a new vehicle, or even a new driver, is the establishment of a baseline setting. Once you have baseline data you can make changes and get faster times. Teams frequently go testing on the same day on the same track, then they watch each other's times for comparison – if they test on different days then they try to factor in changes for the weather.

To enable easy data logging most circuits have one of two systems which can interface with your vehicle.

Stack and *PI* are probably the most popular. The speedometer has a sensor which picks up a signal from the start/finish post. This allows

automatic lap timing and recording in the car. Coupled with a standard built-in trip computer it can give detailed information about performance.

For Kart and other club level events each competitor is given a transponder which sends a signal to the control office (usually in a high tower). This allows the organiser to plot the lap times of each car, usually this gets a special prize or points.

To log the performance of the vehicle in terms of acceleration, braking, fuel consumption, engine operation and cornering forces it can be fitted with sensors to record this information onto a memory chip in a small ECU like device – a data logger unit. This information can be overlaid on the circuit information for full analysis.

The analysis will lead to a series of questions, which, with experience, you will start to develop, for instance:

Could third gear be used at that point?
Why are the tyre temperatures so different?
Is it possible to use more revs out of that corner?

WHOLE VEHICLE SET-UP

Setting up a race car, or bike, should be tackled as a whole. Changing one small part can affect the rest of the vehicle. In this section we will highlight some of the main points for your further research.

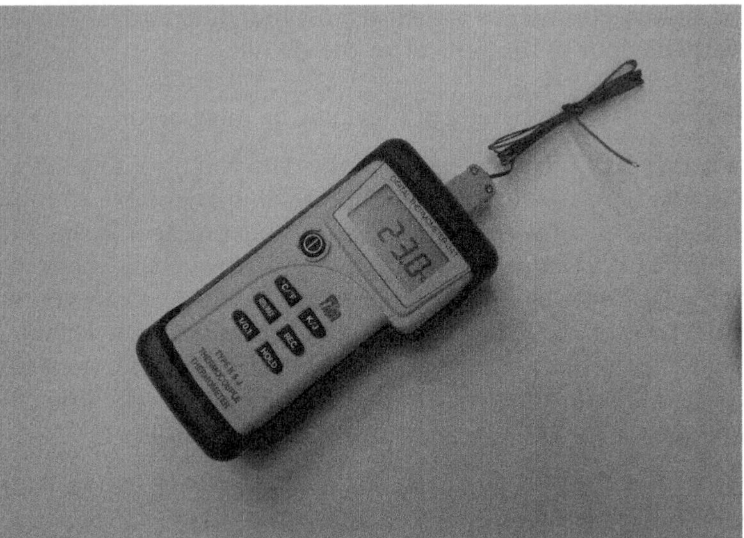

FIGURE 2.15
Digital tyre temperature gauge

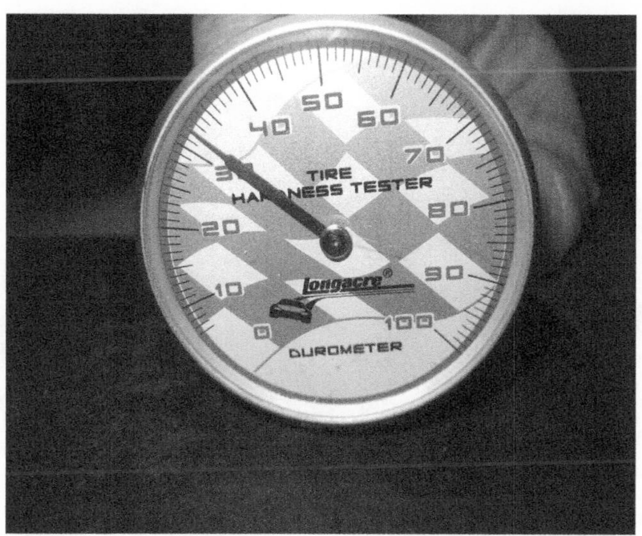

FIGURE 2.16
Durometer for measuring tyre tread hardness

HISTORIC RACER NOTE

The first documented set up of the whole racing car was by William Milliken (USA) and Maurice Olley (UK) funded by Cadillac and under the supervision of Cornell University.

Wheels and tyres – is probably the place to start with any vehicle; the tyres transmit the power, the braking forces and the turning moments between the vehicle and the road. The factors to be considered are:

- rim widths, diameters and off-sets
- different designs and sizes
- tread compounds
- aspect ratio
- front–rear combinations
- different tyre circumferences
- track width
- caster, camber, SAI and toe-in/toe-out.

Centre of gravity (CG) – the aim is to get the centre of gravity into a position which you want it in. Usually it should be as low as you can get it and near the centre of the vehicle. This is measured simply using corners weights; or if you run Formula 1 or WRC you will have a more complex test rig. Standard cars are unlikely to have an accurately placed CG, so getting it accurate may mean moving heavy items like the battery to a different location or adding ballast – lead weights are the normal ballast. This needs to be done both laterally

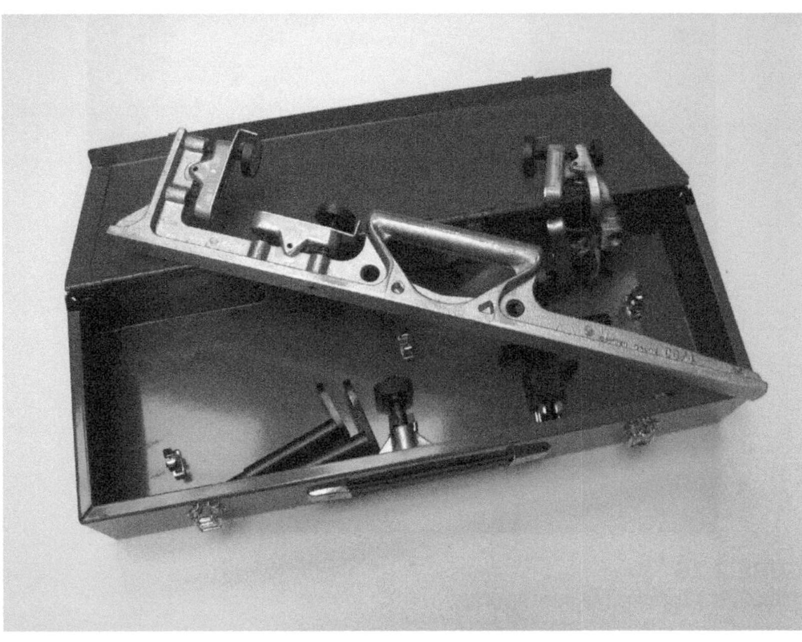

FIGURE 2.17
Dunlop castor, camber and KPI gauge set

and longitudinally. Setting up the static wheel alignment also affects CG, as does ride height.

Ride height and body attitude – the orientation of the body/chassis, also called in aerodynamic terms angle of attack, alters the roll and pitch and yaw. Generally, the more stable position is when the car is higher at the rear than the front and the CG is in front of the centre of pressure (CP). The following points should be considered:

- steering geometry
- bump/roll steer
- anti-dive/anti-squat
- suspension travel
- aerodynamic aids.

Aerodynamic forces and moments – as we have already mentioned these are related to body attitude. In addition you must consider drag, lift, down force and vehicle movements. Often forgotten is the air flow inside and underneath the vehicle. To get this one right you will need to study aerodynamics in detail and use computational fluid dynamics (CFD) software as well as a wind tunnel.

Brakes – vehicle brakes are now well developed and available in various set-ups for most vehicles; but the same problems still need setting up for competition vehicles, these are:

- Cooling – dissipating the heat energy from braking is something which needs careful consideration – the heavier the car the more the heat developed. Consider that if you have a car developing say 250 bhp the heat from the radiator is probably enough to heat an average house. Braking that same car is going to develop the same heat as accelerating it, where is the hot air going?
- Brake balance, or distribution – under heavy braking conditions you will have a lot of weight transfer from the rear to the front, this needs to be managed. On road vehicles the anti-lock braking system (ABS) and possibly the electronic stability programme (ESP) will deal with this. On an open wheel car you are only likely to have two master cylinders and a balance bar – a lot of testing will be needed to get the right choice of brake pads and balance bar position.

HISTORIC RACER NOTE

Jaguar won their first Le Mans 24 hour race by having disc brakes which enabled them to lose less time when braking for corners, so they got quicker lap times than the other cars which had drum brakes.

Drive line – adjusting the gear ratios to suit the vehicle, the power curve and the circuit need accurate attention – the choice of the correct gear ratios will become noticeably important to handling under transient accelerating conditions. That is, when cornering you should enter on the best line for a power exit; at this point it is important to have the correct gear ratio to have the correct part of the power band available to give the required acceleration to power the vehicle into the next straight. Remember that you do not want to be changing gear on your exit from the curve; you want the car to be in a straight line for both maximum handling and maximum weight transfer to put the power onto the track. Gear ratios are further discussed in Chapter 3 on transmission.

Springs and dampers – this area also relates to ride height and body attitude as well as load transfer. The factors to be considered here are:

- Suspension type – IFS, IRS, and type of layout, and suspension medium.
- Spring characteristics – these should be calculated in conjunction with the tyre characteristics.
- Damper characteristics.

The suspension layout will control dive and squat and give other characteristics such as the amount of movement relative to wheel movement and bump steer. On single seater cars it is normal to have inboard mounted dampers and springs to reduce the unsprung weight which in turn improves wheel control. On road cars softer

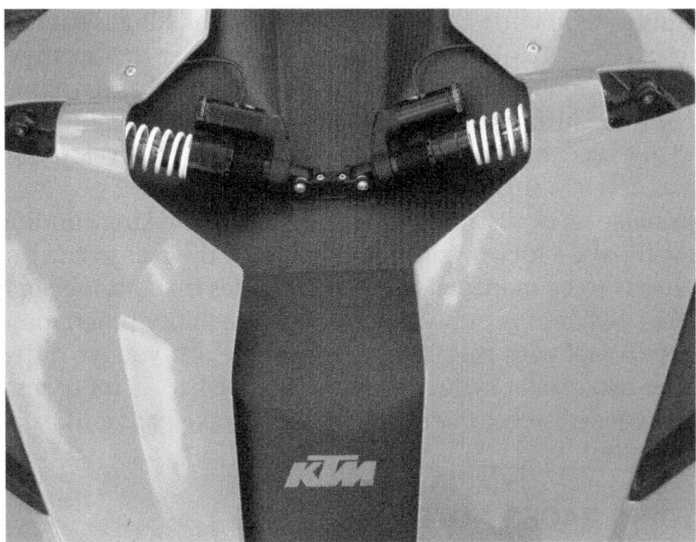

FIGURE 2.18
KTM Horizontal suspension layout

suspension is needed – for rally cars it is normal to have variable rate springs to give some initial movement to cope with uneven surfaces.

Different front to rear spring rates will alter pitch and roll – it is normal to have slightly stiffer springs at the rear than at the front, this keeps the roll centre higher at the rear to enhance stability.

FIGURE 2.19
KTM Front quarter view

FIGURE 2.20
Krugger – look at the chassis on this – has it any possible application on a car?

Compliance – this covers chassis stiffness, suspension stiffness and other factors of body/chassis design. For maximum stiffness many open wheel formula cars use a composite tub construction. Rally cars often incorporate roll over bars to add stiffness as well as safety. Short circuit racers – USA Midgets and Escorts (not to be confused with UK cars of the same name) have stiff tubular steel chassis.

Driver–vehicle interface – often referred to as ergonomics, is probably the most important factor. Does the car suit the driver, or

FIGURE 2.21
Vyrus – look at the suspension on this – is it like any car you know?

does the driver suit the car? The race car engineer can change a number of factors, such as:

- steering gear ratio
- brake pedal force
- clutch pedal movement
- steering feel – kickback and movement
- seat position
- instrument position
- gear lever position – movement.

CHAPTER 3

Transmission

Diagnose and Rectify Motorsport Vehicle Transmission and Drive Line System Faults

IDENTIFICATION

Before working on a transmission it is essential to identify it. The VIN number (see Chapter 2 – 'Chassis') will give basic information about the transmission; but this is usually restricted to transmission type – for example 4-speed automatic, or 6-speed manual. The transmission will have its own identification (ID) tag. The ID tag is usually located so that it can be seen without removing the transmission; it may be stamped on casting lug, a thin plate stuck on a flat surface with glue, or a thicker plate secured with a rivet or screw. The information will show the model year, the transmission type, the model number and the serial number. The numbers will also indicate the year of manufacture, the day of manufacture in **Julian date** format and in some cases the shift time.

The year and Julian date are shown in a four digit sequence. The first digit is the year of the decade, for instance 2009 will just be 9. The next three digits give the day and month in Julian (also called **ordinal**) form. That is 001 is 1 January, 365 is 31 December. So 9219 is 7 August 2009.

CLUTCH AND COUPLING

THINK SAFE

- The dust from clutch linings should not be breathed in.
- Keep clutch dust of your hands.
- Clutch parts get very hot – be aware if working to do an urgent repair.

FIGURE 3.1
Four paddle clutch plate

The function of the clutch is to transmit the engine power to the gearbox allowing a temporary neutral position for the purposes of running the engine with the vehicle at rest and changing gears.

For motor sport applications clutches tend to fall into two categories, **single dry plate** and **multi-plate** clutch packs.

FIGURE 3.2
Six paddle clutch plate

The design of the clutch is controlled by the amount of **torque** it must transmit from the engine to the gearbox. The general clutch calculation stems from the formula:

$$\text{Torque} = \text{No. of Sides} \times \text{Spring Force} \times \text{Coefficient of Friction} \times \text{Mean Radius}$$

This is usually written $T = S\,P\,\mu\,R$ (say torque equals spur). Increasing the value of any of the factors will increase the **torque** (T) that can be transmitted by the clutch. If you are carrying out this calculation use the appropriate units – either Newtons and Metres or Pounds and Feet to give the torque in the units you want – Nm or lbft.

Number of sides (S) – the more friction surfaces in the clutch pack, the greater the amount of torque that can be transmitted – hence the use of multi-plate clutches on race vehicles. Remember a single plate clutch has two sides.

Spring force (P) – is the clamping force of the clutch springs – if this is increased the force required to disengage the clutch will be increased. This means either more effort on the clutch pedal or a different clutch operating ratio. Use either N or lb.

Coefficient of friction (μ) – depends on the clutch friction material. Generally the greater the coefficient of friction the faster the rate of wear, so competition clutches tend to have lower coefficients of friction than road cars. The friction materials fall into a variety of classifications. These are asbestos type (largely banned in Europe but used in the rest of the world); non-asbestos type, this looks like asbestos; ceramic; sintered metallic; and carbon type. Asbestos is banned in the UK because of its likelihood of causing lung damage – asbestosis; non-asbestos uses a mineral alternative – but be aware that all materials which produce particulates (dust) can cause lung damage. The non-asbestos friction materials use a combination of a mineral ore, metal wire or chopped wire and a resin. The construction may be moulded or a woven form. The harder the friction material surface the lower the coefficient of friction. The carbon types use carbon or carbon fibre materials. The carbon fibre material can operate at much higher temperature than the other materials; but the coefficient of friction is much less, also they are much more expensive. Coefficients of friction vary between 0.25 for a carbon clutch, and 0.4 to 0.6 for a non-asbestos type.

Mean radius – this is the average radius of the annular ring of the friction material. The area of friction material affects the wear rate and the rate of heat dissipation. However there is a compromise to be made with these factors. A compromise on race cars is the use of the paddle clutch instead of using a full annular ring. The units for the mean radius are Metres or Feet (not millimetres or inches although these are often used for classification purposes).

Single dry plate – in motor sport applications they are similar in basic appearance and operating principles to a standard clutch used in road cars; but the detail of construction and feel of operation is different. Typical points include:

- Lightweight fly wheel to enable faster engine speed acceleration.
- Lightweight clutch plate and friction disc to enable faster speed change responses.
- Use of paddle shaped friction disc to give best compromise on area of friction material.
- Hard material for wear and heat resistance.
- Strong spring force for maximum clamping force.
- Use of lightweight materials throughout.

Competition clutches tend to be either in or out – that is, the engagement is almost immediate, unlike in a road car where the clutch can be slipped to hold the vehicle, not good practices but sometimes needed in city driving.

Multi-plate clutch – so called because it uses more than one friction disc and hence will transmit much more torque – two friction discs means four friction surfaces, so the torque transmitted by an otherwise equivalent clutch will be doubled.

They are typically available in 2, 3 and 4 plate packs. They are smaller in diameter than a single plate clutch to enable maximum engine speed acceleration. The standard diameters are: 115 mm ($4^1/_2$ inch), 140 mm ($5^1/_2$ inch), 184 mm ($7^1/_2$ inch) and 200 mm ($7^7/_8$ inch).

For some applications multi-plate clutches are run in an oil bath – referred to as wet. The oil bath acts as a lubricant, maintaining an even coefficient of friction between the plates to give smooth operation, taking away dust particles and acting as a coolant. On carbon clutches running them wet actually increases the coefficient of friction.

TRY THIS

Have a look at the different types of clutches in the workshop or on the internet.

Alignment – the alignment to look out for is that of the gearbox input shaft into the crankshaft. The problem with changing engines and gearboxes is that they might not align, especially with adaptor plates; you may need to elongate some of the holes. Also, always use a dedicated clutch aligning tool, not one of the universal type.

TABLE 3.1 Clutch faults

No.	Symptoms	Causes	Rectification	Comments
1	Clutch slip	Greasy facings	Clean or replace facings	
2		Binding pedal/cable/pivot mechanism	Free or replace	
3		Incorrect pedal adjustment	Adjust as needed	
4		Incorrectly re-fitted floor panel or covering	Re-fit correctly	
5	Rattle in Clutch	Damaged driven plate	Replace	
6		Worn release mechanism	Replace	
7		Loose release bearing	Replace	
8		Worn transmission bearing	Replace	
9		Bent or worn splined shaft	Replace	
10		Excessive transmission backlash	Investigate and repair as needed	
11	Fierceness or snatch	Misalignment	Re-align as needed	See text for information
12		Binding pedal mechanism	Free or replace	
13		Worn out or greasy facings	Clean or replace	
14	Clutch judder	Misalignment	Re-align as needed	See text for information
15		Greasy facings	Clean or replace	
16		Mal-adjusted engine tie-bars	Adjust as needed	Check for damage to body / chassis mounting point
17		Uneven wear of friction facings	Replace	
18		Faulty engine mountings	Replace	Rubber can become detached from metal mounting
19		Bent splined shaft or driving plate	Replace	
20	Tick or knock	Worn pilot bearings	Replace	
21		Hub splines worn due to misalignment	Replace	Check alignment
22	Clutch drag (difficult to engage gear)	Greasy or broken facings	Clean or replace	
23		Misalignment	Re-align as needed	
24		Incorrect pedal adjustment	Adjust as needed	Check mechanism completely
25		Sticking driven plate	Clean or replace	
26		Binding/sticking pilot bearing	Free or replace	
27		Damaged pressure plate, driven plate or clutch cover	Repair or replace	Check for cause of damage

Abnormal noises – the first check is to find out whether it is with the clutch pedal up or with it depressed. With the clutch pedal depressed the release bearing will be turning and the friction plate will not. With the clutch pedal up the release bearing will not be turning, the friction plate will be turning clamped inside between the flywheel and the pressure plate.

Vibrations – the same diagnosis applies as the abnormal noises. On race cars balance is very important and the clutch should be balanced along with the crankshaft and other engine parts.

Fluid leaks – these will cause loss of clutch action. On some older race cars the hydraulic clutch may need adjusting.

Slip – see Table 3.1.

Judder – see Table 3.1.

Grab – be aware that race clutches are either in or out, there is no in between.

Failure to release – the first place to check is the operating arm on the clutch housing. On cable systems the most common fault is a broken cable.

TRY THIS

Try driving different cars to get the feel of different set-ups.

GEARBOX

The **manual gearbox** – or **transmission** as it is called, when it includes the differential and final drive, takes the drive from the clutch to the drive shaft or propeller shaft. In doing so it may provide a gear ratio and also alter the direction of the drive.

The choice of gear ratios depends on the vehicle and its application. For example a dragster is likely to use only two gears – you don't have time to make a lot of gear changes in $^1/_8$ mile (200 metres) or $^1/_4$ mile (400 metres). Also the dragster engine has a fairly wide power band. A circuit car may have five or six gears because it has a very narrow power band and a very wide speed range – 30 mph (50 kph) in the esses (bends or chicane) and 180 mph (300 kph) on the straight.

Generally smaller bored engines run at higher speeds in tighter power bands than larger ones; and therefore small engined cars are more affected by the choice of gear ratios.

Power train layout – the relative positions of the engine, gearbox, final drive and drive shafts greatly affects the performance of the vehicle. To get the ideal **weight distribution** the best engine

FIGURE 3.3
Sectioned straight-cut gearbox

position is **mid-engine**. This means that the engine is in front of the transmission – look at Formula Ford cars as an example, this is also used on many Ferraris, Lamborghinis and the MGF. The other advantage for circuit cars is that with the Hewland, and similar gearboxes, the ratios can be changed quickly by removing the complete gear set from the rear and replacing them with another set – you may have a different set for the fast Thruxton circuit as against the slower Indy circuit at Brands Hatch.

FIGURE 3.4
Straight-cut gear set

There is a number of different gear types used: **helical**, **double helical**, **straight-cut** and **epicyclic**. Double helical are mainly found in truck (HGV) gearboxes, epicyclic are used in automatic gearboxes. Helical gears are used in production gearboxes for their longevity, quietness and ability to transmit heavy torque loads with relatively thin profiles.

Competition cars tend to use straight-cut gears as they can be easily made in small scale production relatively cheaply. They can also slide between each other – sliding mesh or when changing a gear range. They are often used with a **dog clutch** to give a positive drive. Straight-cut gears also are noisy in operation and have a limited life – not problems on competition cars.

Abnormal noises – try the gearbox in different gears to see if it is a chipped tooth or a worn bearing. Chipped teeth tend to knock or rattle under power; worn bearings tend to whine on over run. Also look out for loose components and stones or bits of shrubbery caught up in the drive train. Always check the oil level before and after each event.

Vibrations – as for noises try different gears. On motor sport vehicles it is most likely bent drive shafts or damaged drive joints.

Loss of drive – either a slipping clutch or a broken gearbox part.

Difficulty engaging or disengaging – if it is not the clutch it is likely to be:

- worn or damaged synchromesh
- damaged sliding mesh gear
- damaged dog clutch.

TRY THIS

You will learn a lot about gearboxes by driving different cars with different gearboxes, they all have their own feel and characteristics.

AUTOMATIC GEARBOX

The automatic transmission system comprises of closed loop with overriding controls. The system is made up of four main components and the external controls; let's look at each of these in turn. The automatic transmission is operated by the flow of oil, known as automatic transmission fluid. This fluid is very low viscosity and operates at high pressure and high temperature. That is, in excess of 4 bar (60 psi) and 120 degrees Celsius (250 degrees Fahrenheit).

FIGURE 3.5
Gear part detail – sectioned casing

Torque converter – operates using four main components, these are **impeller** (or pump), **stator**, **turbine** (or runner) and the **lock-up-clutch**. The impeller is turned by the engine – what would be the flywheel on a manual transmission. The fluid (automatic transmission fluid, also know as TQF or auto transmission oil) flows from the impeller to the turbine. The turbine is attached to the input shaft of the automatic gearbox. Under certain conditions the fluid is returned from the turbine through the stator to the impeller. This action multiplies the torque between the engine and the transmission. The torque multiplication can be up to four times the engine torque. Remember that slippage is taking place between the engine and the transmission. When the turbine speed reaches 0.8 (80%) of that of the impeller the lock-up clutch is engaged by the oil pressure from the front gearbox pump. At that point the transmission efficiency goes from 80% to 100%. In other words the torque converter is locked so that there is no slippage.

Pump – the automatic gearbox usually has two pumps, a **front pump** and a **rear pump**. The front pump is driven by the torque converter and provides the pressure for the intermediate gear changes. The rear pump is driven by the gearbox output shaft; on fully hydraulic systems its pressure controls the gear shift points. The rear pump takes over from the front pump in top gear after torque converter lock-up. The rear pump works at a lower pressure as it only needs to hold the top gear in place – under these conditions torque is lower than in intermediate gears.

FIGURE 3.6
Gear part detail – input side

Valve block – (a) Controls the flow of the oil for lubrication purposes – oil from the secondary regulator is directed to the bearings and the gears. (b) Distributes the fluid pressure to the torque converter, through the secondary regulator and to the lock-up clutches to engage the gear trains. (c) Regulates the oil pressure for the other functions within the transmission – commonly called line pressure. (d) Activates the lock-up clutches and/or control brake bands to engage the desired gear.

The valve block therefore controls the gear position, depending on the settings of the **gear selector** and the **throttle position** and the **output shaft speed.** It carries out its function through a mixture of mechanical and electronic inputs.

Gear train – the gear train is a series of epicyclic gears, the number of gear sets depending on the number of gear ratios required. By holding the different members of the epicyclic gears it is possible to get different ratios from one gear set.

TRANSMISSION CONTROLS

There are three main controls, namely: **gear selector**, **throttle position indicator** and **output shaft speed**. In addition there are sensors for the transmission oil temperature, the brake operation and the cruise control. The gear selector may be very simple – as on the so called GM *slush boxes* used on drag, sprint or short circuit cars. These use a simple lever to manually select either a single ratio or a limited change of ratio – these boxes only have 3-gear ratios at most. Or they may be very complex offering both manual change and a range of automatic modes. The modes on the automatic transmissions on high performance cars usually offer a sports mode, this means that the engine reaches a higher rev range than the normal – and more fuel economic – mode. Sports mode transmissions usually have an artificial intelligence capability; that is they learn the driver's style and alter the change points to suit the driver. This operates within defined parameters and of course limits the vehicle speed to 156 mph (250 km/h) through the ignition ECU. The automatic transmissions with mode control usually have a separate transmission ECU to control the valve block through a series of actuators.

The throttle position (TP) indicator may be a simple cable control between the accelerator pedal and the valve block; or an electronic control – usually a potentiometer that varies the voltage with the throttle pedal position. Typically the reference voltage to the TP is 5 volts. In throttle closed position the output voltage is 0.5 v whilst at full throttle it is 4.5 v.

The speed sensor is on the output shaft (or tail shaft) and may take a number of different forms, these include: a mechanical governor, a rear hydraulic pump or an electronic sensor. The mechanical governor is used on simple 3-speed gearboxes where the gear changes are at set speeds. With this set up the mechanical TP serves as a kick-down switch – so under part throttle the change from 2nd to 3rd gear may be say 45 mph (72 kph), at full throttle (kick-down) the speed is raised to 65 mph (104 kph). When a rear hydraulic pump is used, under cruising conditions the front pump closes down – the torque converter mechanical clutch will be engaged at this point and the rear pump provides enough pressure to retain the engaged (top) gear, as speed is lost the front pump will be re-engaged and then the drive goes back to the fluid in the torque converter.

The electronic speed sensors may be of the permanent magnet type – as the speed is increased the output voltage increases or a reed switch which closes every time the output shaft rotates – in effect counting the number of rotations.

Brake operation sensor – this may be a connection through the stop light switch. Depending on the make and model the sensor may carry out three functions:

- operate as an inhibitor switch to prevent the engine being started unless the brake pedal is pressed
- disengage the lock-up clutch when the brake pedal is pressed to improve braking by nulling out the engine drive
- Engage low (or 1st) gear early to use engine braking.

Oil temperature sensor – a thermistor is inserted in the valve block to monitor the transmission oil temperature, its signal is passed

FIGURE 3.7
Gear part detail – open casing

to the transmission ECU. The oil temperature is sometimes used to assess the oil viscosity and line pressure may be modified accordingly. When oil temperature rises the ECU may also choose to change up to a higher gear – to reduce the temperature and prevent gearbox damage. A warning light may be invoked at this point.

Cruise control – this is designed to maintain a constant cruising speed under all conditions; it is also programmed on modern cars for maximum fuel economy by monitoring the exhaust gas. The cruise control will change down from 4th gear to 3rd gear if the speed drops below the pre-set figure on a hill. The use of cruise control is very much recommended on motorways; set to 70 mph the legal maximum and the figure above which aerodynamic drag starts to effect fuel economy, it will also give the maximum fuel economy by ensuring the ideal gear change point far more accurately than can be done manually.

Actuators – the actuators are the components which control the flow of fluid to engage the appropriate gear ratio by moving the brake band or clutch. On the older gearboxes the oil pressure is the factor which changes the gear. To put it in basic terms the faster the car goes, the higher the pressure of the transmission fluid (oil) and this changes the gears by the operation of the valves in the valve block. The choice of gears can be overridden by selector and the throttle kick down. On electronic controlled transmissions solenoids move the valves in the valve block according to signals from the transmission ECU. There are two types of solenoids in use – the **on-off type** and the **linear type**. The on-off type as the name suggests simply have two positions; on–off (or open-closed). The principle of operation is like that of a starter solenoid – an armature is moved inside a magnetic coil against a spring. That means when no current is applied the solenoid is off and the valve closed by spring pressure (always remember basic engineering spring close, cam open), when an electric current is applied to the coil the armature moves to open the valve. The linear, or plunger, solenoid operates like an electronic petrol injector. That is, it pulses between open and closed according to a signal from the ECU. The signal is sent in the form of a low voltage and variable current duty ratio pulse signal – to give it its full name, usually referred to as duty ratio. This can be likened to squeezing the end of a water hose pipe with your fingers – you can pulse the water out in an infinite number of different ways, the length of each pulse controls the volume of water and hence how wet things get. Therefore the linear solenoid allows the valve to be opened different amounts and therefore gear changes can be smoother as the gear is engaged more gently.

The duty ratio is the percentage of time that the power is applied over a cycle – measured as a percentage (%). The duty cycle is the length of time between the start and the finish of the cycle – measured in milliseconds (ms).

Shift timing – The timing of the gear shift is very important to both performance and economy. The older fully hydraulic systems change up and down depending on the vehicle speed, the gear selector position and the throttle position. The gear changes on these gearboxes tend to be abrupt and fuel consumption is higher than

FIGURE 3.8
Gear part detail – drive output

with a manual gearbox because of the limitations of only having three gear ratios and some slip of the torque converter.

On electronically controlled gearboxes the shifts are controlled by the ECU. The ECU is fed information by the gear selector, the throttle position, the engine speed, the vehicle speed, the engine load and the gearbox temperature. The transmission ECU may also be linked to the engine management system ECU. The purpose of this link is to smooth the gear shifts. For example on an upshift (1 to 2, or 2 to 3) the transmission ECU sends a reduce torque signal to the engine ECU; the engine ECU responds by reducing the fuel supply to the engine, so reducing engine torque whilst the shift is carried out. This is carried out in about 3 or 4 milliseconds. For racing motorcycle engines running in the 15,000 to 20,000 rpm bracket a micro-switch is fitted to the manual gear change pedal, so that on upshifts the fuel supply is cut in the same way. During downshifting the change of gear is often smoothed by retarding the ignition timing.

The electronic control gives improved performance and economy and elongates the life of the transmission by: allowing the use of lower line pressure; reducing slippage of friction parts by smoothing the shifts which in turn reduces the operating temperature and keeps the transmission oil clean.

Fault diagnosis – before starting to try to find any fault on an automatic transmission carry out the following three tasks:

1. Read the manual – make sure that you know exactly how the transmission works.
2. Write down the exact symptoms (fault) described by the driver.
3. Check that the symptoms described by the driver are correct – not all transmissions have the same functions – driver modes – and what a driver thinks is a fault can simply be an absence of that function on a different model.

THINK SAFE

Never remove, or dismantle, any automatic transmission until you are certain what is wrong with it – it is almost impossible to visually identify faulty components unless the fault is obvious.

With the gearbox in situ the first checks, in order of operation, are shown in Table 3.2

After carrying out the checks 1 to 6 you may want to confirm your thoughts or make a few more checks if nothing is obvious. The next two tests are a **stall test** and **pressure testing sequence**. We will look at each of these tests in turn.

TABLE 3.2 Basic Automatic Transmission Checks

No.	Item	Look for	Comment	Next step
1	Casing, seals, bearing housings	Oil leaks and signs of impact damage		Repair or replace as needed
2	Transmission fluid	• Check level • Check condition – it should be clear and odour free	If oil is 'brown' or smells of 'burning' it is likely to be a burnt out clutch or brake band	Likely to need a complete replacement of clutch or brake bands and mating surfaces
3	Cable controls	Fraying, lack of adjustment or sticking	This will affect kick-down	Rectify as needed
4	Electrical controls	• Loose connections • Damaged components • Broken cables	Visual inspection of items	Rectify as needed
5	Warning lights and switches	Check the operation of lights and switches	Often drivers do not understand the function of lights and switches	Use findings to inform further diagnosis
6	Onboard diagnostics	Check for fault codes	If a fault code is shown this must be dealt with before proceeding further	

Stall test – don't try this on old vehicles, you may risk destroying the gearbox completely.

1. The engine must be fitted with a tachometer (rev counter), or a portable one is connected to the engine so that you can read the engine speed.
2. Chock the wheels and apply the brakes.
3. Engage 'D' and apply full throttle for no more than 30 seconds.
4. Note the speed at which the engine stalls, compare to manufacturer's data – typically this is between 1200 and 1600 rpm. If the stall speed is:
   - too low then check for torque converter slip, or low engine power
   - too high then check for slipping of the front clutch, or low line pressure.

Pressure test – this will vary on the vehicle, so follow the manufacturer's instructions.

THINK SAFE

Pressure testing is best conducted on a rolling road, never on a public road, check your pressure gauges for accuracy, and wear appropriate PPE – including goggles and gloves to give protection from hot fluid under pressure. You should have a full risk assessment for this activity.

Gearboxes have different tapping point; however most of them have a general point to take line pressure.

Typical line pressure is between 4 bar and 7 bar (60 psi to 110 psi). In top gear it is likely to be lower than when in intermediate gear using the pump to hold gear only. Table 3.3 shows typical faults from the line pressure test readings.

Abnormal noises – check whether these are under drive, or idle, or over run situations. Check the obvious first – loose coupling, mountings and heat shields. On rally cars look for stones and branches being caught in the transmission.

Vibrations – these are very much like the abnormal noises – the most usual causes on motorsport vehicles are worn couplings or bent shafts.

Loss of drive – see Tables 3.2 and 3.3.

Failure to engage – see Tables 3.2 and 3.3.

Failure to disengage – see Tables 3.2 and 3.3.

Leaks – see Tables 3.2 and 3.3.

Failure to operate – see Tables 3.2 and 3.3.

TABLE 3.3 Typical Line Pressure Reading Faults

No.	Reading		Likely fault
1	Low at idle in all positions		Low oil, restricted filter, oil leak – internal or external, sticking open relief valve
2	High at idle in all positions		Stuck closed main regulator, faulty electrical control
3	Low only in one of the following	N	valve body
4		R	low / reverse servo or valve body
5		P	low / reverse servo
6		1	forward clutch, low / reverse valve body, or servo
7		2	forward or intermediate clutch, or valve body
8		D	forward clutch, or valve body

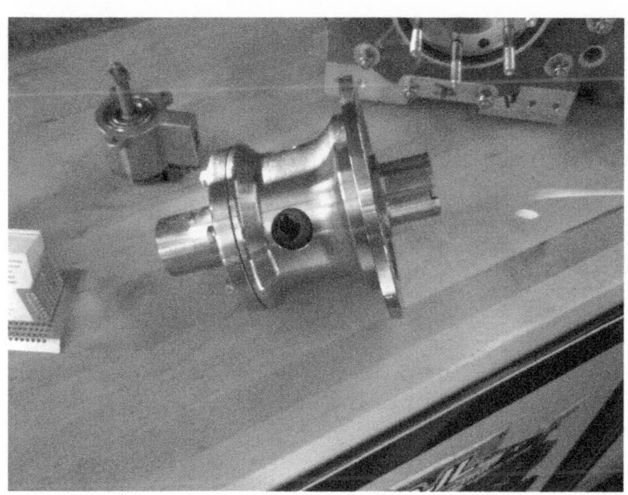

FIGURE 3.9
Gear part detail – LCD differential

Incorrect shift patterns – see Tables 3.2 and 3.3.

Electrical and electronic faults – see Tables 3.2 and 3.3.

FINAL DRIVE

The final drive may be incorporated with the gearbox for a transmission unit or separate as a rear or front axle. Open wheel cars tend to have the transmission behind the engine – using a combined gearbox and final drive.

On front engined, rear wheel drive cars, using an open propeller shaft, the rear axle may be a **live axle** or a **dead axle** with separate drive shafts or a version of a **swing axle**. With any of these layouts, it is normal to change the final drive ratio to suit the circuit or type of event. Changing the rear axle ratio changes the **overall gear ratio** (OGR) without the need to change the gearbox ratio.

$$\text{Overall gear ratio} = \text{Gearbox Ratio} \times \text{Final Drive Ratio}.$$

The Overall Gear Ratio, and hence the Final Drive Ratio is calculated to meet the needs of the vehicle in the motorsport event in question. For instance faster circuits will need a higher top speed than slower circuits.

HISTORIC RACER NOTE

Up until the 1970s it was normal for a *clubman* driver to drive the car to events – that is not to use a trailer, as is current practice. Also the clubman may use the same car for a number of different events, such as hill climbs, circuit events and road rallies, changing wheels and tyres and gear ratios to suit. You can see such vehicles at the Historic and Classic Shows.

Setting gear ratios – the gear ratios are chosen to match the engine's performance characteristics to the requirements of the event, this may include:

- top speed to meet the engine's maximum power
- hill climbing performance
- acceleration characteristics
- slow speed driving.

The main factors to be considered when calculating gear ratios are:

Air resistance (R_a) – this is the product of the drag coefficient (C_d), the surface area of the front of the vehicle (A) and the square of the velocity (V^2)

$$R_a = C_d \, A \, V^2$$

As you can see, the air resistance increases with the square of the velocity (speed in a given direction), so the faster the vehicle goes the greater the increase (exponentially). As a general rule air resistance becomes important above 70 mph (112 k/h) and as vehicles are rarely travelling directly into a wind, think of all the corners on a circuit. Instead of taking the straight frontal area most aerodynamicists base their calculations on an elevation of the vehicle taken at 14 degrees to one side.

Gross vehicle weight (GVW or just W) – this is the weight of the vehicle with driver, load and full of fuel and oil.

FIGURE 3.10
Differential, and crown wheel and pinion

Rolling resistance (R_r) – is the resistance of the vehicle to roll along a surface, this is mainly due to tyre wall and tread deformation. The rolling resistance is the product of the weight of the vehicle (W) and a coefficient (a coefficient is a number which does not have units and is usually found by experimentation). The **coefficient of rolling resistance** (C_r) varies with the tyre tread, the tyre wall design, and the road, or track surface.

$$R_r = C_r W$$

HISTORIC RACER NOTE

The aspect ratio of tyres has reduced over the years, on pre-WW 2 cars it was 100%, then up until the 1970s 80% was normal, then 60% and now less than 50%. To give wider wheels and tyres, to improve cornering, the wheel diameters were first reduced. Then to give the same rolling radius with the low profile tyres the wheel diameters have been increased; so that rolling resistance has reduced from a figure of 0.30 to less than 0.01.

Linear wheel velocity – is equal to **linear vehicle velocity**. That is if the car is doing 70 mph (112 k/h) then the tyre tread is doing the same speed. For gear ratio purposes we need to convert this into **wheel revolutions** – revolutions per minute (rpm) or easier to work with revolutions per second (rev/s), another alternative is radians per second (rad/s).

To calculate the wheel speed in rev/s: take the road speed (convert to miles per second or metres per second) and divide it by 2π times the **rolling radius** (r_r) of the tyre.

$$\text{Circumference} = 2\pi\, r_r$$

The rolling radius is the radius from the centre of the wheel to the ground under load, this figure changes with the load on the tyre and the tyre pressure. Under race conditions the tyre pressure increases with increase in temperature; it also reduces as the tread is worn.

Gradient resistance (R_g) – is the force needed to overcome the incline. A down hill is a negative resistance (in other words positive). The easiest way to deal with a gradient (G) is to find the sine (sin) of the gradient.

$$R_g = \sin G\, W$$

Tractive effort (TE) – is the force exerted by the tyre on the road to propel the vehicle forwards; that is the wheel torque divided by the rolling radius. The tractive effort must exceed the sum of the resisting forces for the vehicle to move forwards. That is:

$$\text{TE must be} > R_a + R_r + R_g$$

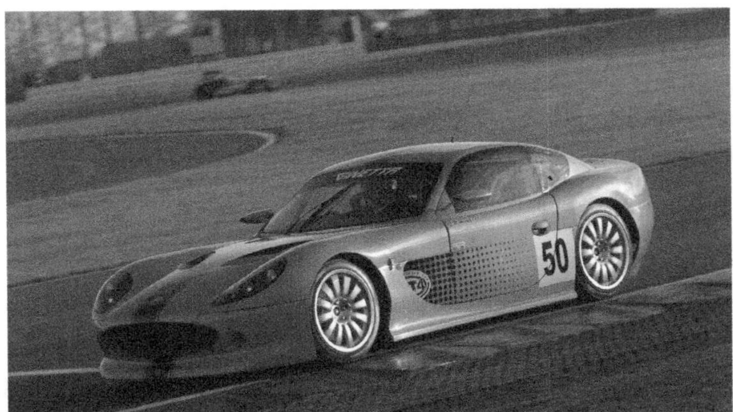

FIGURE 3.11
Ginetta

If you know the engine power (bhp or PS or kW) and the speed at which it occurs then you can calculate the rear wheel torque and hence TE.

Top gear is calculated by finding the road speed at which the TE will equate with the total resistance ($R_a + R_r + R_g$).

Bottom gear is calculated by using the engine torque figure at the speed at which the clutch is normally fully engaged to find the TE which is greater than the total resistance.

RACER NOTE

On motorcycle engined cars, such as the Yamaha R1 engined Legend, or the Hyabusa engined Westfield, which have power bands starting at about 8,000 rpm, you must let the clutch up gently as you move off from rest otherwise the engine will stall as the engine develops its power from high revs rather than torque. Once the vehicle is moving there is no problem – it is just overcoming the inertia of rest in addition to the other resistances.

Abnormal noises – wear in the final drive is usually indicated by a whining noise – this may increase on over run, let the accelerator (gas) pedal up on a down hill to check. Clicking noises are either bearing failure – light click; or damaged tooth, heavy click on power.

Vibrations – the two main reasons are damaged drive shafts or worn joints.

Loss of drive – check for broken joints.

Oil leaks – check oil seals and gaskets first. On rally cars, or circuit cars which have had an off, check for hair-line fractures of the casing.

Fitting sump guards to rally cars reduces the risk of damaged gear casings – these may be extended to complete skid plates. Formula open wheelers often use a fibre board to do a similar job – they may also tie in with aerodynamic control regulations.

Failure to operate – this is usually due to broken gears, or internal drive shafts, either through fatigue or misuse, such as dropping a gear, or engaging reverse by mistake.

DRIVE LINES AND COUPLINGS

Motorsport vehicles have a variety of drive lines and couplings not found on standard road cars. In this section we will discuss some of these and the problems related to them.

The power and torque transmitted by drive lines and couplings on competition vehicles is both high in ultimate figure and liable to be applied rapidly – with hammer force.

CASE STUDY

The Bentley Azure T

No.	Item	Figure
	Power	500 BHP/507 PS/373 kW
	Torque	738 lbft/1000 Nm
	Acceleration	0–60 mph 5.2 s/0–100 km/h 5.5 s
	Top speed	180 mph/300 km/h

If this car had an overall gear ratio (OGR) of 3 to 1. The torque on the crown wheel would be 3,000 Nm, that is:

New Torque = Torque × OGR

1,000 Nm × 3 = 3,000 Nm

If the crown wheel were 0.25 m in diameter the force would be:

Force = Torque/Radius

3,000/0.125 = 24,000 N

If this force is applied quickly, for instance equal to the acceleration of gravity, we use the 2σ (say two sigma) rule, that is the force is doubled, becoming 48,000 N (approximately 10,000 lbft).

CHAPTER 4

Inspection

Inspect Motorsport Vehicles

INTRODUCTION

This unit builds on the content of the other motorsport units and it is suggested that it is studied concurrently with them. The unit is laid out for the student to use in conjunction with the IMI Data Collection Sheets, with suggested notes for each sheet data collection sheets. The choice of vehicle to inspect will depend on your specific area of interest, Kart, Single Seater, Saloon or Rally Car; and the availability of vehicles. The wider the range of vehicles that you can use when training, the wider and more varied will be your experience and employability. Of course you may choose to specialise at an early stage. Both approaches have advantages.

What is important is that you carry out vehicle inspections and record your findings. Carry out inspection with the IMI Data Sheets which you can obtain direct from the IMI if you are registered, or may be supplied by your tutor. If you have not got access to these IMI Data Sheets then you may use MOT or service check sheets as guidance.

SAFETY FIRST

When approaching a vehicle for the first time, especially a damaged motorsport vehicle, you must carry out a RISK ASSESSMENT. That is a mental assessment of the situation. If you are on duty at an event, may be as a team mechanic, or a marshal, you will be first on the scene. In this case you need to consider the following:

- Is it safe to get near the vehicle? Think about location, other traffic, other people – you must put your safety as first priority.
- The next priorities are making the scene safe, calling for help, and the application of First Aid and perhaps the Paramedics.

If you are approaching the vehicle after its recovery, even when it is in the pits or workshop, a mental RISK ASSESSMENT is needed. Some typical examples are:

- The rally mechanic who opened the bonnet of a vehicle at the end of a forest stage to be scalded by the coolant from a radiator fracture – the coolant was at about 130 degrees C.
- The NASCAR mechanic who decided to touch the brake disc when the car came in the pits for a wheel change, he burnt his fingers, the disc was probably at about 800 degrees C – remember that you need to wear gloves to even touch the tyres.
- The technician who was feeling under the seat to plug in the diagnostic connector and had his skin pierced by a hypodermic needle left by a previous occupant.

USE YOUR SENSES

When inspecting a vehicle it is always good to use your senses, but do it with care.

TABLE 4.1

Sense	Checks	Cautions
Sight	How does this vehicle look? Is it level and square? Are there any leaks or stains? Signs of damage or misuse?	Use eye protection
Sound	How does this vehicle sound? Listen to the different systems or parts.	Use ear protectors
Smell	Is there a smell which might indicate a leakage or overheating?	Wear a mask
Taste		Not advised
Touch	Use your finger tips to check for damage or wear. Use a nail to check whether a blemish is raised or sunken.	Use hand protection
Kinaesthetic	Feel the operation of controls or mechanical linkages for smoothness.	Be prepared for the unexpected

RESPECT FOR VEHICLES

As a technician YOU ARE RESPONSIBLE for the vehicle which you are inspecting, therefor you must NOT cause damage to the vehicle, even if the vehicle which you are inspecting is seriously damaged. You never know what repairs or salvage may take place. You are expected at all times to:

- use seat covers
- use floor mat protectors
- use wing covers
- protect from bad weather
- jack-up and support the vehicle safely using appropriate jacking points
- ensure that the systems are treated with care
- remove finger marks.

FIGURE 4.1
Scrutineering garage at Lydden Race Circuit

FIGURE 4.2
Scrutineering underway at Lydden Race Circuit

FIGURE 4.3
Compression gauge in use

FIGURE 4.4
Ignition timing light

Task 1: Inspect Vehicles – Tyre Markings and Wheel Serviceability

THE CRITERIA (see Table 4.2)

(This is what you need to do and know to complete the unit, completing the Data Collection Sheet is evidence of this.)

TABLE 4.2

No.	Task	Detail	Typical answers	Action points
1	Use necessary PPE	Mechanic's gloves	Protects hands from cuts and burns when handling tyres	Look out for hot tyres and sharp edges
2	Correctly identify relevant tyre data for vehicle	Size, pressure, type, fitment	175/65 R 14 2 bar (30 psi) M&S symmetrical	Use either vehicle mfg or tyre mfg data sheets
3	Locate and identify tyre markings	Tyre and wheel size, speed rating, load index, tread wear indicator, aspect ratio	175/65 R 14 82 H	Sketch tread wear indicator between tread ribs
4	Measure tread depth	Use MOT green depth gauge	Legal minimum 1.6 mm	Check across full width of tread and all circumference
5	Examine tyre condition	Look for damage to the tyre	Cuts, lumps, bulges, tears, abrasions, intrusions, movement on rim, concussion, tread separation	Check the tyre pressure
6	Examine tread wear patterns	Look for uneven wear (see 7), skid wear and patch wear on tread	Skidding, out of balance	
7	Identify reasons for abnormal wear patterns	State causes of edge or wear at one point	Incorrect tyre pressures, steering misalignment	
8	Examine wheel condition	Look for damage to wheel	Impact damage, cracks, distortion, run-out, security	
9	Check valve condition and alignment	Is it in straight?	May be damaged or bent after impact	
10	Record faults			List wheel and tyre faults
11	Complete Data Collection Sheet			Record findings on Data Collection Sheet

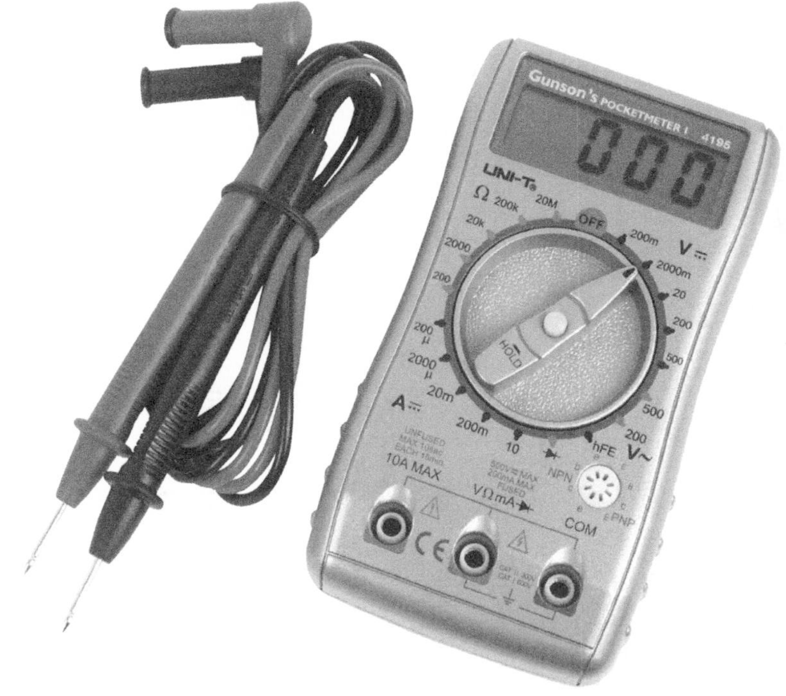

FIGURE 4.5
Multimeter

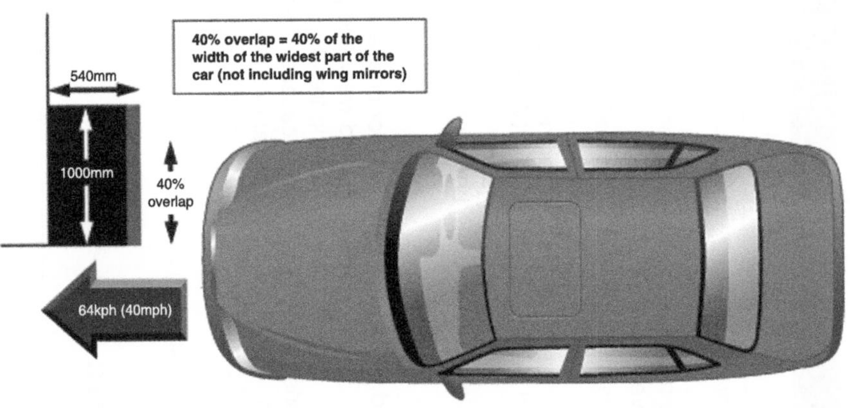

FIGURE 4.6
Frontal impact test

Task 2: Inspect Vehicles – Braking System Serviceability

TABLE 4.3

No.	Task	Detail	Typical answers	Action points
1	Use necessary PPE	Mechanic's gloves, goggles		Remember brake fluid damages paint work and brake parts get very hot
2	Check serviceability of components	Measure against manufacturer's data	Record thickness of lining at each service point	
3	Check lines and hoses	Corrosion, deterioration, leaks	Service records	On race cars these should be replaced each season
4	Check security of components, lines and cables	Lines, hand brake cable, ABS/wear indicator wires		Check when steering/ suspension is moved
5	Check for pressure retention and leaks	Check fluid level, visual check	Test brake fluid for water content	Hold pedal down for 30 seconds with engine running for servo operation, it should not move
6	Correctly adjust free play	Hand brake and foot brake	MoT regulations	Are wheels free to spin when brakes are released?
7	Test brake servo for operation	Is servo operating?		Test brake pedal feel with and without engine running
8	Check drum/ disc for serviceability	Check for wear, ovality/ run-out	Oval drums, distorted discs	Use dial gauge indicator
9	Check warning devices and on-board diagnostics		Diagnostic print out	
10	Check brake light operation		Do they stick on when brake is released?	
11	Record faults			Record any braking system faults
12	Complete Data Collection Sheet			Record findings on Data Collection Sheet

Task 3: Inspect Vehicles – Suspension System Serviceability

TABLE 4.4

No.	Task	Detail	Typical answers	Action points
1	Use necessary PPE	Mechanic's gloves, goggles		Use of four-post MoT specification hoist is recommended
2	Visual inspection of ride height	Use manufacturer's data	Record height at each corner	Use plastic tape to avoid damage to paint work
3	Check security of suspension components, mountings and fixings	Components, chassis mounting points, locking devices, adjacent vehicle structure	Floor or other mounting points may be distorted following an off	
4	Examine springs	Security, corrosion, cracks, fractures, wear, play	If removed from vehicle check spring rate (kg/inch)	
5	Check hydraulic/ pneumatic system	Fluid retention, leaks	If topped up record volume	
6	Check suspension members for wear/damage	Wishbones, tie rods, radius arms, panhard rods	Look for accident damage	Compare to new parts if needed
7	Check for free play in suspension joints	Follow MoT guidelines	Worn or loose joints	Use wheel free and lever as appropriate
8	Check CV/ UJs for wear/ damage		Torn boots, wear	Check on full lock both left and right
9	Check suspension warning devices and on-board diagnostics		Diagnostic print out	
10	Record faults			Record any suspension system faults
11	Complete Data Collection Sheet			Record findings on Data Collection Sheet

Task 4: Inspect Vehicles – Steering System Serviceability

TABLE 4.5

No.	Task	Detail	Typical answers	Action points
1	Use necessary PPE	Mechanic's gloves, goggles		Remember brake fluid damages paint work and brake parts get very hot
2	Check column/ rack/box/idler mounting		Corroded chassis mountings and loose mounting bolts	Get a colleague to move the steering wheel whilst you look
3	Check for free play	Check manufacturer's data	Old cars may have an inch of free play at the steering wheel, new ones have no free play at all	Each system is different; older classic racing cars often have a lot of free play and require an experienced feel for adjustment – ask your tutor
4	Visual inspection of gaiters/ dust covers		Check for oil or grease leakage	Often you need to feel the gaiter, or pull it a little
5	Check for fluid retention and leaks		Look for oil marks on adjacent components	As above
6	Check steering system components	Rack, box, idler, track rod, TRE, column, wheel, joints, collapsible joints		
7	Check behaviour lock to lock	Check steering components and wheels for fouling	Tyres fouling on inner wings on full lock	
8	Check steering damper	Test operation and look for leaks	Oil leaks	You may need to remove one end mounting to do this
9	Check wiring		Loose or fouling	
10	Check warning devices and on-board diagnostics		Diagnostic print out	
11	Record faults			Record any steering system faults
12	Complete Data Collection Sheet			Record findings on Data Collection Sheet

Task 5: Inspect Vehicles – Lighting System Serviceability

TABLE 4.6

No.	Task	Detail	Typical answers	Action points
1	Use necessary PPE	Mechanic's gloves, goggles		
2	Check side lights	Colour, position and operation	Look at lighting regulations	Remember to include the number plate light
3	Check rear fog lights	Should only operate with ignition and headlights on		Check dash board warning light
4	Check headlights	Should be matched pair with appropriate dip		
5	Check headlight main beam aim	Check against MoT and manufacturer's recommendations	Record amount of dip	
6	Check stop lights	Check for operation and colour	Often loss of colour in lens; if bad earth side light may dim when brake lights applied	Check that they go out when the brake is released
7	Check reflectors	Check position and operation		
8	Check indicator lights	Check for colour and operation – check flash rate	Check with all lights on to ensure good earth	Check dashboard warning light
9	Check hazard lights	Check operation		Check dashboard warning light
10	Record faults			Record any lighting system faults
11	Complete Data Collection Sheet			Record findings on Data Collection Sheet

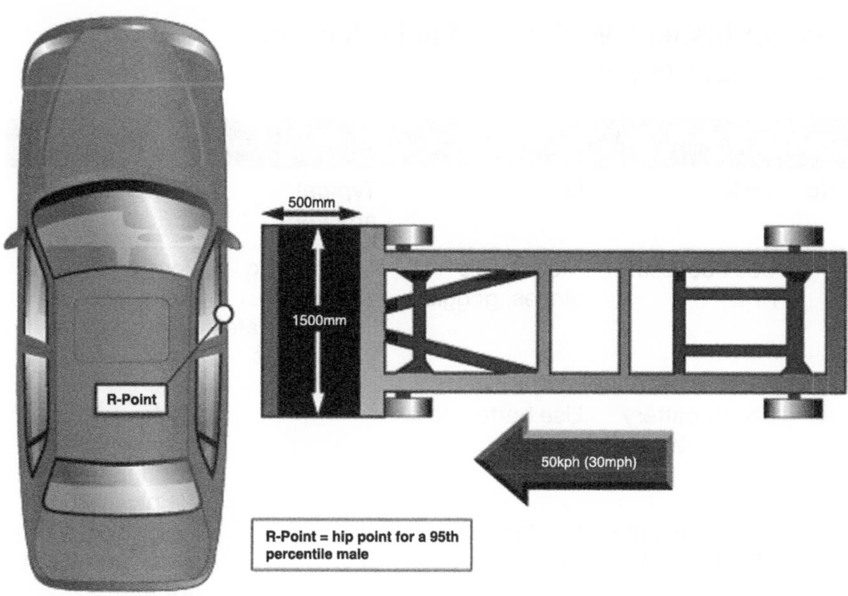

FIGURE 4.7
Side impact test

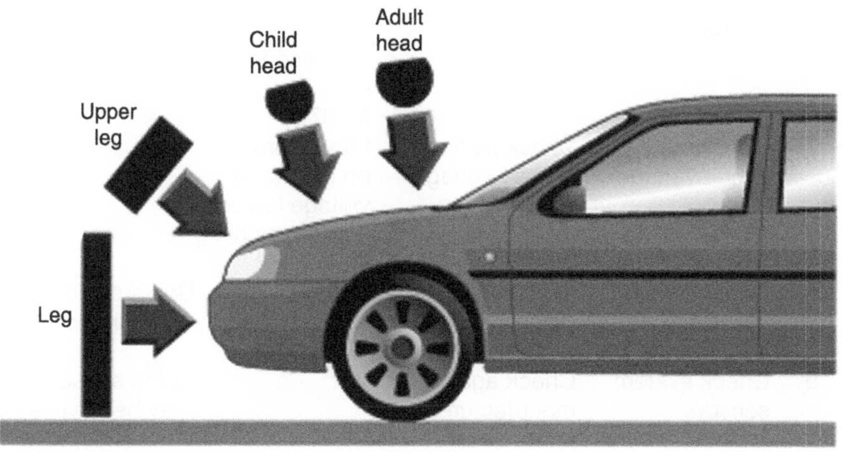

FIGURE 4.8
Pedestrian impact test

Task 6: Inspect Vehicles – Fuel Injection System Diagnosis

TABLE 4.7

No.	Task	Detail	Typical answers	Action points
1	Use necessary PPE	Mechanic's gloves, goggles	No smoking or naked light, use safe low voltage lighting	Remember petrol is highly flammable
2	Check battery serviceability	Use battery test procedure		System will not operate without a sound power supply
3	Check security of connections	Electrical and fuel lines		Remember the filter – and has it got fuel
4	Check diagnostic codes		Diagnostic print out	
5	Check fuel pressure on fuel rail	Check manufacturer's data	Typically 60 to 100 psi (4 to 7 bar)	
6	Check injector operation	Dwell opening period	Typically 3 to 7 milliseconds (0.003 to 0.007 s)	
7	Check supply circuit voltage	Check against battery voltage	May operate on regulated voltage lower than nominal 12 v	
8	Check circuit continuity	Compare resistance readings – ohms		Be careful of damaging circuit
9	Check system sensors	Check against manufacturer's information		Some sensors may be covered in other systems
10	Record faults			Record any system faults
11	Complete Data Collection Sheet			Record findings on Data Collection Sheet

Task 7: Inspect Vehicles – Ignition System Serviceability

TABLE 4.8

No.	Task	Detail	Typical answers	Action points
1	Use necessary PPE	Mechanic's gloves, goggles		Remember HT can be over 40kV – this could kill you
2	Check battery serviceability	Use battery test procedure		System will not operate without a sound power supply
3	Check connections for security	LT and HT		Visual check for HT voltage leakage – tracking
4	Check supply voltage	To LT components	May be less than nominal battery voltage	
5	Check coil resistance	Use manufacturer's data	Compare resistance in ohms	
6	Check HT output voltage	Use manufacturer's data		
7	Check spark plug condition	Also confirm type from application data	Use spark plug fault finding chart	
8	Check HT lead resistance	Use manufacturer's data		
9	Check supply and earth continuity		Ohms	
10	Check supply to pulse generator		May be less than nominal battery voltage	
11	Check output of pulse generator	Use manufacturer's data		
12	Record faults			Record any system faults
13	Complete Data Collection Sheet			Record findings on Data Collection Sheet

Task 8: Inspect Vehicles – Manual Transmission System – Clutch

TABLE 4.9

No.	Task	Detail	Typical answers	Action points
1	Use necessary PPE	Mechanic's gloves, goggles, dust mask		Old clutches in classic cars may contain asbestos; in any case avoid breathing the dust and keep it away from your clothes
2	Determine symptoms		Typically: slip, grab or non-disengagement	
3	Carry out visual inspection	Cables, hoses, cylinders, pipes and linkages	Cable clutches often suffer broken damaged cables; hydraulic ones tend to suffer fluid leaks	Check fluid level on hydraulic system first
4	Check free play adjustment	Check manufacturer's data		Check for free play at operating lever end as well as pedal
5	Check for transmission oil leaks		Look for oil drips or stains	Check transmission oil level
6	Check clutch assembly for wear or damage	Lining thickness and plate condition	Wear/cracks/heat damage/spring strength/general deterioration	
7	Check release bearing		Should run freely and quietly	
8	Check seal and gaskets		Check for leaks	
9	Check clutch assembly and mountings for damage or corrosion	Gearbox mountings/ pedal box/ cables stops/ line joints		Especially cable clutches that tend to break through their bulk head mountings
10	Record faults			Record any system faults
11	Complete Data Collection Sheet			Record findings on Data Collection Sheet

Task 9: Inspect Vehicles – Lighting System Diagnosis

TABLE 4.10

No.	Task	Detail	Typical answers	Action points
1	Use necessary PPE	Mechanic's gloves, goggles		Be careful when testing the circuits, disconnect the battery before disconnecting any wiring
2	Check components and wiring for security		Cables correctly bound or taped, cable ties tight	Especially under wheel arches
3	Check battery serviceability	Use battery test procedure		System will not operate without a sound power supply
4	Check fuses for serviceability	Condition, connections, ratings, continuity	Check with manufacturer's data	Remember, too large a fuse rating will not give sufficient protection
5	Check bulbs for serviceability	Condition, connections, ratings, continuity		Replace bulbs at regular intervals
6	Check relays for serviceability	Operation, feed voltage, output voltage	Check against design settings	Take special care with solid state relays, do not test with high voltage or current
7	Check supply to switches	Check manufacturer's data		May be less than nominal battery voltage
8	Check supply from switch		Should be same as supply voltage	
9	Check supply to lights		Check volt drop	
10	Check earth connections	Continuity and volt drop	Check volt drop and resistance	
11	Record faults			Record any system faults
12	Complete Data Collection Sheet			Record findings on Data Collection Sheet

Task 10: Inspect Vehicles – Auxiliary System Diagnosis

TABLE 4.11

No.	Task	Detail	Typical answers	Action points
1	Use necessary PPE	Mechanic's gloves, goggles		
2	Check auxiliary system component/ wiring for security and serviceability	Condition, connections, ratings, continuity		
3	Check battery serviceability	Use battery test procedure		System will not operate without a sound power supply
4	Check fuses for continuity	Condition, connections, ratings, continuity		
5	Check motors and components for serviceability	Operation and output		
6	Check relays for serviceability	Operation, feed voltage, output voltage	Check against design settings	Take special care with solid state relays, do not test with high voltage or current
7	Check supply voltage to switch			May be less than nominal battery voltage
8	Check output from switch		Same as input	
9	Check supply to components		Volt drop and resistance	
10	Check earth connections	Continuity and volt drop	Volt drop and resistance	
11	Record faults			Record any system faults
12	Complete Data Collection Sheet			Record findings on Data Collection Sheet

Task 11: Inspect Vehicles – Cooling System Diagnosis

TABLE 4.12

No.	Task	Detail	Typical answers	Action points
1	Use necessary PPE	Mechanic's gloves, goggles		Remember coolant is both scalding hot and under pressure until the engine cools
2	Determine vehicle mal-operation symptoms	Questioning customer/ supervisor	Overheating, overcooling, coolant loss	
3	Visual check for security and serviceability		Hoses loose, chaffing on other cables	
4	Check coolant level and anti-freeze content	Use appropriate coolant hydrometer, check manufacturer's data	Typical 50% anti-freeze giving protection to –15 degrees C	Many systems only use pre-mixed coolant, this should be changed regularly
5	Carry out pressure test on cooling system		Not holding pressure, coolant leak	Take care with electric fans
6	Carry out pressure test and vacuum test on cap	Check with manufacturer's data	Pressure: 15 psi and vacuum: 4 inHg	Some systems have more than one cap – these may not be sealed ones
7	Check radiator for serviceability and security		Detached/broken mountings, grille broken or clogged with dirt (especially rally cars)	
8	Test thermostat operation	Observe both opening and closing	Typically open at 88 degrees C, check for full closure	Opening temperature is usually stamped on the thermostat
9	Check cooling fan operation	Drive may be: electrical, fluid coupling or direct drive		Take care – don't do this if you have any form of epilepsy
10	Check temperature gauge operation	Compare with similar vehicle or instrument type	Should operate in middle range under normal conditions	
11	Record faults			Record any cooling system faults
12	Complete Data Collection Sheet			Record findings on Data Collection Sheet

Task 12: Inspect Vehicles – Starting System Diagnosis

TABLE 4.13

No.	Task	Detail	Typical answers	Action points
1	Use necessary PPE	Mechanic's gloves, goggles		
2	Determine vehicle mal-operation symptoms	Questioning customer/ supervisor		
3	Check battery serviceability	Use battery test procedure	Volt drop figure	System will not operate without a sound power supply
4	Check starting system for security and serviceability	Battery terminals, starter and solenoid terminals, loose cables	Corrosion on battery terminal	Flat battery or loose/corroded terminals account for about 30% of all AA and RAC non-start call outs
5	Check battery voltage	Check volt drop on starting	Should be > 10.5 volts	
6	Check supply to starter solenoid	Check volt drop and resistance of feed cable	Approximately battery voltage	
7	Check supply voltage to starter	Check volt drop and resistance between solenoid and starter motor	Approximately battery voltage	With pre-engaged starter motor this will not apply
8	Check volt drop on supply circuit	Combine 6 and 7 above		
9	Check solenoid contact volt drop			
10	Check earth circuit volt drop	Volt drop and resistance between battery and chassis and engine and chassis		Particularly check battery earth lead to chassis and if there is an earth strap between the engine and the chassis – these are often lost when an engine is changed or replaced
11	Record faults			Record any system faults
12	Complete Data Collection Sheet			Record findings on Data Collection Sheet

Task 13: Inspect Vehicles – Lambda Sensor Diagnosis

TABLE 4.14

No.	Task	Detail	Typical answers	Action points
1	Use necessary PPE	Mechanic's gloves, goggles		
2	Determine vehicle mal-operation symptoms	Questioning customer/ supervisor		
3	Check battery serviceability	Use battery test procedure		System will not operate without a sound power supply
4	Visually inspect lambda sensor/ components/ wiring for security and serviceability		Loose connections and corrosion	
5	Check engine at normal operating temperature	Check against typical models	Use probe in engine oil if appropriate	Best if car is run for 4 miles or more
6	Check voltage at normal operating idle	May use oscilloscope	Check against manufacturer's data	
7	Check voltage change with rich air : fuel mixture			
8	Check voltage supply to sensor			May vary from battery voltage
9	Check sensor earth continuity		Should be zero	
10	Check sensor resistance			
11	Record faults		Diagnostic print out if possible	Record any system faults
12	Complete Data Collection Sheet			Record findings on Data Collection Sheet

Task 14: Inspect Vehicles – Lubrication System Diagnosis

TABLE 4.15

No.	Task	Detail	Typical answers	Action points
1	Use necessary PPE	Mechanic's gloves, goggles		Remember engine oil runs at over 100 degrees C, often 150 to 180 degrees C on a circuit car
2	Determine vehicle mal-operation symptoms	Questioning customer/ supervisor		
3	Visual check lubrication system	Oil level, seals and gaskets	Leaks, loss of pressure, oil filter not recently changed – dirty/rusty	
4	Check oil level and signs of contamination	Oil level – exact when settled, colour or dirt	Black – needs changing, brown – usually sign of water contamination – blown head gasket	Engine oil on a race/rally vehicle should be clean at all times
5	Check air filter for oil contamination	Remove and dismantle air filter	Oil in air cleaner indicates breather fault of back firing	No filter on race vehicle – check around inlet tract
6	Check oil pressure	Use an accurate pressure gauge	See 10 below	Remove warning light switch to screw in gauge
7	Check supply to oil pressure switch	Is switch and main gallery clean?		Use calibrated accurate gauge to be certain
8	Check oil pressure warning light	Compare to gauge	Standard typically on at 5 psi, high value ones on at 15 psi	High value switches are available for competition cars
9	Check crankcase ventilation system	Clean, secure and in good condition		Engine will slow when crankcase is vented to open
10	Check smoke emission		Should be no visible smoke	Blue smoke indicates burning oil

No.	Task	Detail	Typical answers	Action points
11	Record faults		Use a square table to show pressure when cold / hot and idle / cruising speeds Oil pressure <table><tr><td></td><td>Cold</td><td>Hot</td></tr><tr><td>Idle</td><td></td><td></td></tr><tr><td>4,000 rpm</td><td></td><td></td></tr></table>	Record any system faults
12	Complete Data Collection Sheet			Record findings on Data Collection Sheet

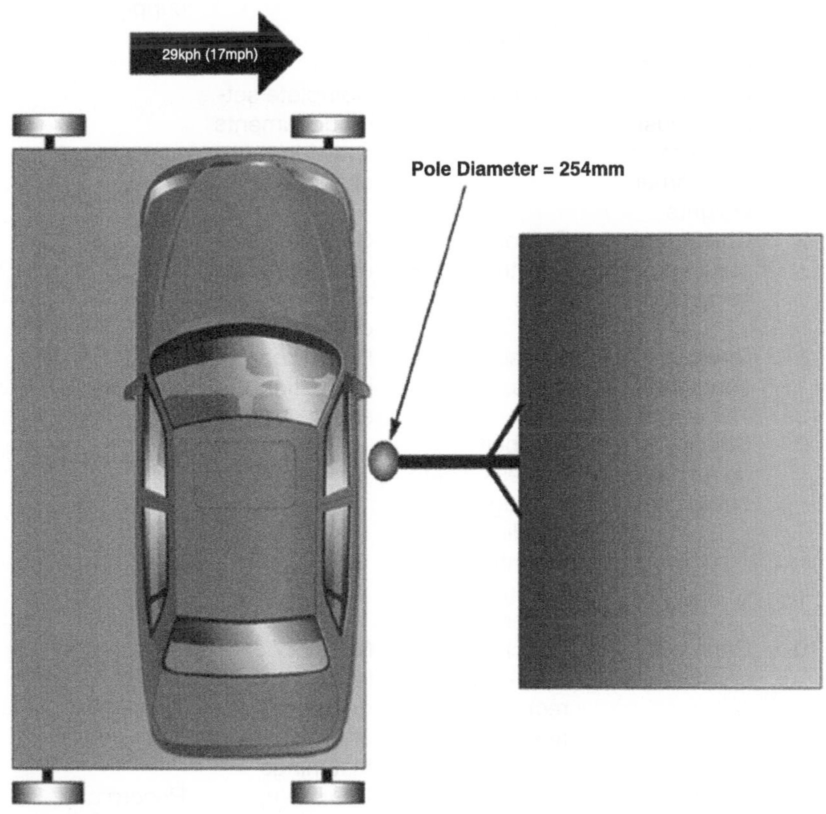

FIGURE 4.9
Pole Test

Task 15: Inspect Vehicles – Pre-Event Set Up - Workshop

TABLE 4.16

No.	Task	Detail	Typical answers	Action points
1	Use necessary PPE	Mechanic's gloves, goggles		
2	Obtain vehicle set-up data	Check with appropriate authority	Use set-up management system	This may need a signature
3	Check previous data and driver comments		Use set-up data log	
4	Check battery serviceability	Use battery test procedure		System will not operate without a sound power supply
5	Check oil and fluid levels	Note variance		
6	Check and adjust suspension and corner weights	Refer to data	Complete set-up documents showing changes	
7	Check and adjust steering geometry	Castor, camber, SAI, toe-out on turns	Complete set-up documents showing changes	
8	Check brake conditions	Measure pads, discs	Record on inspection sheet	
9	Check driver safety equipment	Seatbelt harness, seats, roll cage, padding and mountings, fire extinguisher system		
10	Carry out spanner check	Check against check list; rectify as needed	Note any loose, damaged, or broken fastenings	
11	Record faults		Use set-up data file	Record any system faults
12	Complete Data Collection Sheet			Record findings on Data Collection Sheet

Task 16: Inspect Vehicles – Modification – Electrical System

TABLE 4.17

No.	Task	Detail	Typical answers	Action points
1	Use necessary PPE	Mechanic's gloves, goggles		
2	Isolate electrical system	Disconnect battery		Remember that competition cars often have a more complex battery connection system – especially single seaters with connections for an external starting battery
3	Identify modification instructions/ data	Break this down into stages if appropriate		
4	Modify system to plan		Photograph modifications	
5	Visual inspection of modifications	Visually check all cables and connections	Use check list	
6	Check all connections and relays for serviceability	Supply voltage and feed to circuit voltage		
7	Check power supply to switch		Battery voltage	Fully check before re-connecting power supply
8	Check power supply from switch		Battery voltage	
9	Check modified circuit voltages		As per instructions/ data	
10	Check earth connections	Continuity/ volt drop	Should be zero volt drop	
11	Record faults			Record any system faults
12	Complete Data Collection Sheet			Record findings on Data Collection Sheet

Task 17: Inspect Vehicles – Modification – Fuel System

TABLE 4.18

No.	Task	Detail	Typical answers	Action points
1	Use necessary PPE	Mechanic's gloves, goggles		Remember petrol is highly flammable
2	Isolate electrical system	Disconnect battery		Remember that competition cars often have a more complex battery connection system – especially single seaters with connections for an external starting battery
3	Identify modification instructions/data	Break this down into stages if appropriate		
4	De-pressurise fuel system	Use appropriate container		
5	Carry out modification to plan		Photograph modifications	
6	Visual inspection of modifications for serviceability and security			
7	Check fuel system	Leaks, pressure retention and cleanliness		
8	Check injector wiring	Security, installation and continuity		
9	Check fuel rail pressure and injector operation	Use appropriate data source	Record findings in data file	
10	Record faults			Record any system faults
11	Complete Data Collection Sheet			Record findings on Data Collection Sheet

Task 18: Inspect Vehicles – Modification – Lighting System

TABLE 4.19

No.	Task	Detail	Typical answers	Action points
1	Use necessary PPE	Mechanic's gloves, goggles		
2	Isolate electrical system	Disconnect battery		Remember that competition cars often have a more complex battery connection system – especially single seaters with connections for an external starting battery
3	Identify modification instructions/ data	Break this down into stages if appropriate		
4	Modify system to plan		Photograph modifications	
5	Carry out visual inspection	Check for compliance with appropriate rules and regulations		MoT regulations for road going cars, MSA Blue book for all competition cars, specific regulations for event or series
6	Check for security and serviceability	Connections, relays, supply and feed voltages		
7	Check supply to switch		Battery voltage	
8	Check supply from switch		Battery voltage	
9	Check earth connections	Continuity/ volt drop	Should be zero volt drop	
10	Record faults			Record any system faults
11	Complete Data Collection Sheet			Record findings on Data Collection Sheet

Task 19: Inspect Vehicles – Pre-Event Inspection – At Event

TABLE 4.20

No.	Task	Detail	Typical answers	Action points
1	Use necessary PPE	Mechanic's gloves, goggles		
2	Check previous data and driver comments		Use set-up data log	Check previous data and driver comments
3	Check tyres	Fitment (direction/corner), pressure, type, compound	Use set-up data log	
4	Check oil and fluid levels	Note variance	Check oil and fluid levels	Note variance
5	Check and adjust steering geometry	Castor, camber, SAI, toe-out on turns	Complete set-up documents showing changes	
6	Check security of body work	Panel attached, clips closed, screen clear, decal correctly located	Use check list	May choose to photograph for record
7	Check seat belt harness with driver in place	May need adjusting		Note any pre-sets
8	Arm fire extinguisher system	Check setting, switches		
9	Record faults		Use data log	Record any system faults
10	Complete Data Collection Sheet			Record findings on Data Collection Sheet

Task 20: Inspect Vehicles – On Event Inspection

TABLE 4.21

No.	Task	Detail	Typical answers	Action points
1	Use necessary PPE	Mechanic's gloves, goggles		Remember vehicle components and systems remain hot for a while at the end of a stage or round
2	Check new data and driver comments	Download data logger and record verbal comments	Use set-up data log	Check previous data and driver comments
3	Raise vehicle and remove wheels	Visual inspection		
4	Inspect wheels and tyres for wear/damage		Record wear findings	
5	Carry out spanner check	Check against check list; rectify as needed	Note any loose, damaged, or broken fastenings	
6	Refit wheels	Check security and bearings		
7	Check fluid levels and pressures		Complete check sheet	
8	Check security of body work	Panel attached, clips closed, screen clear, decal correctly located	Use check list	May choose to photograph for record
9	Refuel vehicle	Follow team procedure	Record fuel taken	Remember that petrol is highly flammable
10	Record faults		Save data file	Record any system faults
11	Complete Data Collection Sheet			Record findings on Data Collection Sheet

Task 21: Inspect Vehicles – Post-Event Inspection

TABLE 4.22

No.	Task	Detail	Typical answers	Action points
1	Use necessary PPE	Mechanic's gloves, goggles		Accurate recording and signing-off against names is recommended
2	Obtain and collate race data	Set-up sheets, data logging, fuel and tyre records		There may be both electronic and paper logging systems
3	Analyse race data	Record comments and make to-do list	Jobs to be completed before next event	
4	Check and correct log records	Team de-brief and post-event analysis	Include data check and other comments	
5	Clean vehicle and secure for transporting	Check for damage and fit transportation tyres	Photograph for record	
6	Drain fuel if appropriate	Use safe storage procedure		Measure fuel left for fuel consumption calculations
7	Remove and charge battery			Gel batteries must be slow charged
8	Complete data records	Data logging and vehicle records	Complete all data logging and set-up sheets	Secure storage of data is essential – back up all files and keep in secure place
9	Complete Data Collection Sheet			Record findings on Data Collection Sheet

CHAPTER 5

Overhaul

Overhaul Mechanical Units

This is a generic unit – notice the 'G' in the reference code. That means that the tests relating to this unit will be of a general nature – not motorsport specific. In other words, the sort of jobs that you could do on any vehicle: car, motorcycle or truck. As with all generic units it should be studied concurrently with the area-specific work.

Before carrying out any overhaul work you should check the following:

- The customer fully understands the cost implications.
- You have access to any special tools or equipment which are needed.
- Spare parts are available.
- Repair, set up and test data is available.

COSTS

The cost of overhauling a component – such as a gearbox – can, in some cases, be greater than that of a new one. On the other hand, given that not many parts require replacement, the overhaul may be one tenth of the new item. The biggest factor is often the labour cost, followed by the cost of obtaining the part needed. So the labour charge-out rate is a very big factor.

Let's look at a possible example using a labour rate of £85 per hour.

TABLE 5.1 Overhaul cost comparisons

Overhaul costs		Replacement costs	
Item	£	Item	£
Remove and re-fit gearbox 1.5 hours	127.50	Remove and re-fit gearbox 1.5 hours	127.50
Overhaul gearbox 5 hours	425	Replacement gearbox	500
Parts and oil	175	Oil	15
VAT	127.31	VAT	112.43
Total	854.81	Total	754.93

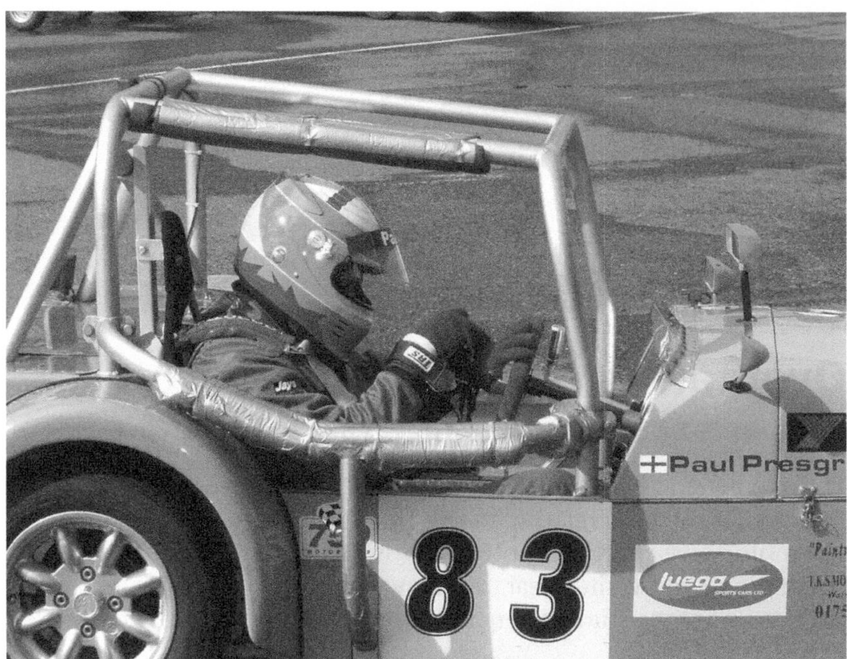

FIGURE 5.1
Paulio racing Hicost

In addition the replacement gearbox will have a period of warranty attached – giving the customer peace of mind.

When it comes to replacement parts such as engines and gearboxes the customer may wish to have the original part overhauled rather than fit a replacement simply to maintain the originality of the vehicle. Rather like the blacksmith who said: 'This is the original hammer that I've had since I was an apprentice – it has had five new heads and six new shafts.'

Beware, however, that when buying or selling units there are many different definitions, or uses in this field, for example:

Original Parts – OEM (original equipment manufacturer) – the units used on the vehicle when it was new. These are often not actually manufactured by the vehicle manufacturer – but sourced from an approved supplier. The boxes may have the logo of the vehicle manufacturer to show approval – but exactly the same item without the logo on the box may be obtainable from the local motor parts factors for 50% of the price. Also, be aware that many manufacturers – particularly German ones – often use re-cycled parts or products made from re-cycled materials. Such parts are still by definition OEM and will carry a price premium.

Re-manufactured units – particularly engines and gearboxes – the components are completely stripped, inspected, measured and crack

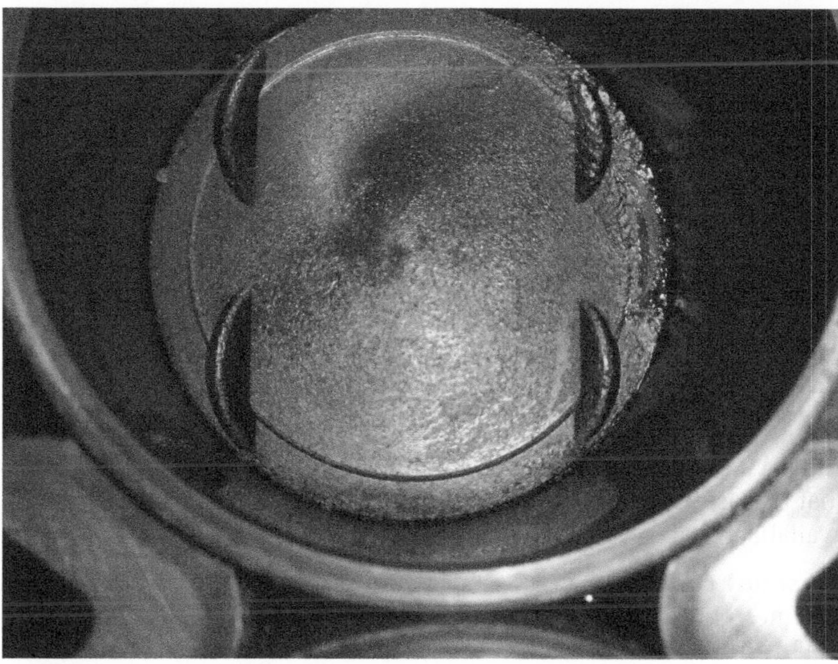

FIGURE 5.2
Piston crown

tested; then all the machined faces are re-machined and new wearing parts fitted – for example bearings and pistons.

Overhauled units – in this case the units are stripped and cleaned and inspected; but only the parts which are outside tolerance are replaced.

Pattern parts – these are look-a-like items which fit exactly and do the same job as the OEM part but are made by another company. Sometimes the pattern parts perform better than the originals and often they are made by the same company which makes the originals but sold under a different brand name, usually these units are 30% cheaper than the OEM equivalent.

HEALTH, SAFETY AND THE ENVIRONMENT

Before carrying out any work whatsoever, you must ensure that you fully understand the Health and Safety issues and are able to comply with them along with the relevant Environmental Regulations and codes of practice. As a re-cap of your work at Level 2 you should consider:

Acts and Regulations – Heath and Safety at Work Act, COSHH, EPA.

Personal Protection Equipment – overalls, safety footwear, gloves, goggles, masks.

Vehicle Protection Equipment – wing covers, seat covers, floor mats.

Identifying System Hazards – such as fuel and high-voltage sparks.

Safe Disposal – using correct oil/fluid drainers, returning displaced units, disposing of electronic and pyrotechnic items correctly.

Fire Hazards and Safety – knowing the muster point, especially if you are at a different site, such as the pit lane at a circuit. Evacuation drills and fire extinguisher identification.

Procedures for dealing with accidents – knowing the contact persons and radio calls if at an event, or on a different site.

Risk Assessments – before carrying out any task you must ensure that risk assessments are in place and the COSHH leaflets are available for all substances to be used.

Personal conduct – you MUST conduct yourself professionally at all times, there is a code of conduct for this if you are a Member of the Institute of the Motor Industry, see: www.motor.org.uk

It must always be borne in mind that *motorsport can be dangerous* and that we are always under detailed scrutiny by other people.

WORKSHOP

To be able to carry out any form of overhaul or repair – apart from the running repairs during competition, some form of workshop is needed. It sounds obvious, but often the functions of the workshop are forgotten, so let's have a look at some of the reasons for a workshop:

- Provide safety, security and protection from the weather for the vehicle.
- Provide safety, security and protection from the weather for the engineers and other staff.
- Provide safe and secure storage for the tools, equipment and spares.
- Provide somewhere to work on the vehicle.
- Provide somewhere to overhaul and repair the units and components.
- Provide an office facility.

Workshops come in all sorts of shapes and sizes. I have worked in them varying between a lock-up garage in a block without a permanent power supply, and a Formula 1 team in a three-storey futuristic building with an underground entrance using solar power and heat-pumps to give a zero-carbon footprint. The ideal workshop is one where the race car can flow through in one direction. This should fit in with the loading and unloading of the car transporter. Preferably having some form of docking-station so that the race car is kept both secure and dry on both its outwards and inwards journeys.

The race vehicle workshop is much more than a garage in the fact that it has a wide range of activities and a larger number of departments. One company I was involved with included a museum, car sales and mail order. We are not discussing these areas in this chapter. If you are employed in the motorsport industry, you may be involved in all aspects of motorsport vehicle overhaul, or just one tiny part. No matter how small, it will be an important function and it will help if you have the bigger picture of what goes on in other departments. Some motorsport companies do just one of the functions covered in the next few pages.

Cleaning down area – when entering into the building there will be a cleaning, or wash, area. Depending on the sort of competition involved this will vary. Obviously a 4 × 4 off-road vehicle will have more mud to remove than a Formula Ford – unless maybe the FF has had an off. This area may also be used for removing body panels – especially damaged ones and other items such as sump guards and changing the wheels to ones more suited for the workshop.

Environmental Protection Act (EPA) and associated Local Authority (LA) Regulations and Building Regulations (Building Regs) require that all buildings comply with certain criteria. The cleaning down area must have drainage traps so that contaminated water, mainly meaning contaminated by oil and chemicals, is not allowed to enter the main drainage system. It is normal to re-cycle the cleaning down water, using a filtration and treatment system. This is environmentally clean and saves the water bill.

Preparation area – an area of bays where each vehicle can be worked on individually – the layout of the bays and the equipment in them will depend on the type of vehicle. For single seaters it is normal to have stands or trestles so that the vehicle is at an ergonomically sound working height. In other cases wheel-free or four-post hoists (ramps) may be used. For a racing team it is usual practice to have a designated bay for each car and a designated technician too. In which case the bay will be kitted out with that technician's tools and the equipment and spares appropriate to the vehicle. In a jobbing shop – where cars are prepared for a variety of owners and/or drivers – the bay will be the sanctity of the technician – cars moving in and out as needed.

In the case of major unit overhaul, the bay may be used to remove the unit only, and then the vehicle moved to a compound area for secure storage until the unit is ready to be re-installed.

Machine shop – this is where the machine tools and similar are laid out and used. Race cars, even touring car classes, tend to use a number of bespoke parts – ones made especially for that vehicle. So the machine shop is needed for both manufacture and overhaul. Also this may be where a number of special tools are kept – the ones which require floor mounting – as against those which can be taken to the benches in the preparation bays. Typically a machine shop may contain some, or all of the tools in Table 5.2.

Bench work area – the bench work area is often around the outside of the vehicle bays, using metal topped benches with drawers and cupboards underneath. Vices and other tools may be mounted on the benches.

TABLE 5.2 Machine shop equipment

No.	Item	Specification	Purpose	Comment
1	Small lathe	6 to 8 inch (150 to 200 mm) swing with 18 to 24 inch (450 to 600 mm) between centres	Making small items such as spacers, and cleaning up round parts	This will need a range of tools and chucks
2	Off-hand grinder – small	Approximately 6 inch (150 mm) diameter wheels – 1 fine, 1 coarse	General sharpening and cleaning	
3	Off-hand grinder – large	Approximately 10 inch (250 mm) diameter wheels	Sharpening drills and tools	Keep flat for accurate work
	Pillar drill	5/8 inch (15 mm) chuck and variable speeds	Variety of drilling	Need variety of vices, or clamps, and drills and countersinks
4	Band saw	Approximately 4 to 6 inch (100 to 150 mm) cut	Cutting up steel stock	
5	Hydraulic press	20 ton (20 tonne)	Removing and replacing bearings and pins	
6	Milling machine	5-axis CNC milling centre	Manufacturing small parts	Used in conjunction with CAD system
7	Buffing – polishing	Floor mounted buffing and polishing heads	Finishing parts	
8	Parts cleaning bath	Chemical cleaning bath with pressure spray	Cleaning parts	

Where the work is solely unit based – such as overhauling gearboxes – the benches may be aligned in rows separate from the vehicles – with the use of stands, or rigs, for the gearboxes or other major components. The drawers, or open racks, will then contain the special tools needed for the job in hand.

Fabrication area – this is where items are fabricated and welded. Usually contains rollers, bending machine, croppers and MIG or TIG welding equipment. Specialist trained engineers in this area will provide these services to enable you to carry out your overhaul and repair tasks.

Composites shop – where composite components are manufactured or repaired – a specialist clean area. This shop is staffed by specialist technicians who will make new parts, or carry out specialist repairs to enable you to overhaul the motorsport vehicle.

Design studio – where vehicles and components are designed and modified. Usually using Computer Aided Engineering (CAE) – that is the Computer Aided Design (CAD) is connected to the Computer Aided Manufacture (CAM) machine tools such as the 5-axis milling centre. You will find the design staff supportive in providing technical data for your overhaul procedures.

Model shop – named because they make models, or macquettes, of vehicles for design and testing purposes – including scale models for the wind tunnel – as well as specialist full scale parts such as aerofoil wings. The model shop is both a source of data and specialist parts and skills. The model shop in larger, or older, firms may incorporate clay and/or wood handling equipment and skills.

Paint shop – where the vehicles are painted. Again this shop has specialist staff and equipment. The overhauled vehicle, or its panels, will be re-finished in this shop. An interesting move in re-finishing is to use vinyl film instead of paint – the US Army dragster – which is probably the fastest race car ever made – uses vinyl film. They applied this in their workshop at the Indianapolis Raceway, known as the *Brickyard*. The comment of the technician applying the material – which is printed and cut on site, was, "It's lighter than paint." When it is on it is almost impossible to tell that it is not paint.

Parts and storage – the safe and secure storage of parts, both new ones, and ones waiting for completion of the overhaul, is very important. They can easily go missing. Small parts have an attraction for the floor (its called gravity – 9.81 m/s^2) then they roll behind the largest possible box so that you can't find them. That aside, race car parts cost 10 or 100 times the equivalent of the road-going equivalent. And some people just want to have the damaged piston out of the number 12 car that didn't win last race of the season; or similar.

Let's have a look at this in a bit more detail. You need a secure storage area large enough to store the large parts and a set of drawers and trays for storing the smaller parts. When you are stripping a component – such as a gearbox – then you need a tray system laid out to keep all the small parts – nuts, bolts, washers, spacers and so on in order so that they can be assembled in the reverse order. If you are doing repetitive work – such as overhauling gearboxes, then you will probably be provided with suitably marked out trays – you will also be able to identify each component without any thought.

To ensure absolute recognition there are a number of procedures, two frequently used ones are:

- Plastic bags – like food storage bags – put each component into a bag and label the bag with the part name, part number, customer details (car number or VIN) and any other details – such as *left rear*. The bags are then placed in a tray which also acts as customer, car and unit identification.
- Photographs – take photographs of both the assembled and the stripped parts. Record the photograph numbers, most digital cameras do this automatically, and reference the numbers to your notes.

These two techniques are used extensively in specialist firms – such as when repairing, or overhauling, Ferraris. Cars like Ferraris may fall into a model category, such as F430; but customers may have upgraded the seats, engine specification and other details; or the originals may have been replaced following an accident, with used parts from another model. As it is unlikely that a workshop manual exists, you will rely on the parts you have – so look after them, make notes and take photographs.

You will also need a storage system for every-day consumables, that is items which are used on a day-to-day basis in the repair or overall of vehicles or units. Such items may include:

- cleaning cloths
- polishes and detergents
- hand cleaning materials
- specialist cleaning solutions
- screws, nuts and bolts
- specialist fixings – such as toggle fasters
- washers, spacers and shims
- locking wire
- wiring cable
- electrical connectors and fasteners
- tape and adhesives
- gaskets and seals.

These items may be charged out against jobs in two ways:

1. individually against the job number
2. on each customer's bill as either a percentage of the bill or pro rata against the number of hours worked.

Dynamometer and test shop – if working on complete vehicles you will need to have access to a rolling road dynamometer – referred to as the dyno, or the rolling road, by most staff. This allows the road wheels to sit on the dyno rollers and the power and torque measurements taken. Running any competition engine makes a lot of noise – so this is usually situated separate form the main building, or suitably noise insulated from it.

If it is just the engine which is being tested, then a test cell is used. A test cell, of course, requires much less space than a rolling road. This takes the form of the engine mounted on a frame which is attached to the dynamometer (dyno). Because of their exceptional noise, aircraft engines are tested underground; some race teams do this too.

Checking and loading area – this is a secure area for the prepared race vehicle to be kept safely prior to, and during, loading onto the transporter.

Office – office space is needed for a number of functions in motorsport, these include:

- Reception – to meet and greet customers and to record customer information and carry out tracking of customer's repairs.
- Finance and administration – to control the flow of cash and communications with customers and suppliers.
- Meeting area for the team and directors – often this takes the form of a board room with a big table and chairs that can be used for a number of functions.
- Engineering office area – with a racing team the race engineers will need an area to work on data from testing and racing, that is, to analyse the data and work out strategies for future developments. This usually takes the form of bar type desking where the laptop computer can be docked into the team network – usually still hard-wired for security, though high-level encryption may be used with a suitable wire-free system (WiFi). This is a separate area to that of the design team who are working on upgrades and new designs for future vehicles and components.

In addition to the functional part of the activities in the office area this area can be used for displaying trophies – *silverware* – and photographs. Therefore the security of the office must be considered with appropriate locks and an alarm system. Interestingly, many of the Formula 1 teams have private museums at their Headquarter (HQ) offices – those of McLaren and Williams are outstanding, forming a detailed history with examples of each type of car used.

TOOLS AND EQUIPMENT

Table 5.2 lists the equipment needed for the Machine Shop. In this section we will look at the fuller range of equipment which is used in vehicle and unit overhaul.

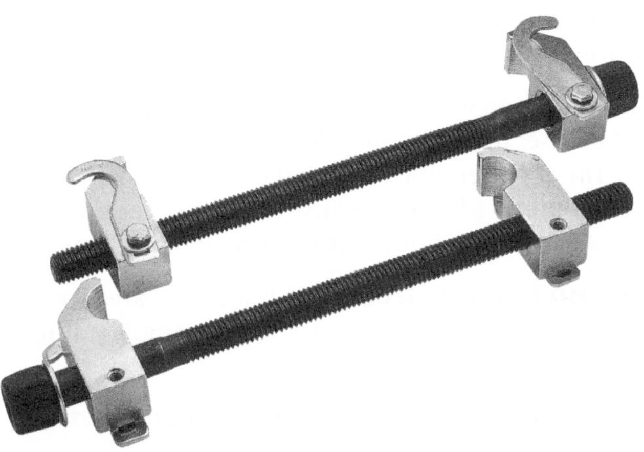

FIGURE 5.3
Spring coil compressors

FIGURE 5.4
Ball joint splitter

FIGURE 5.5
Impact sockets

FIGURE 5.6
Ratchet spanners

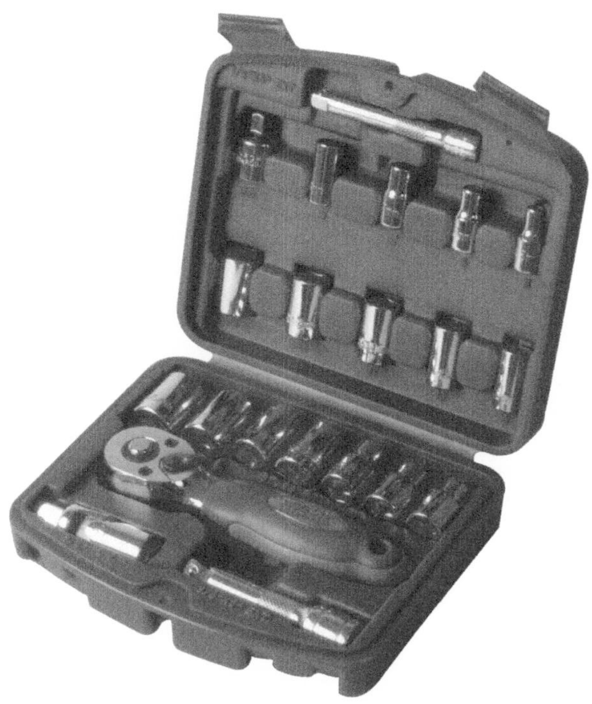

FIGURE 5.7
Socket set

FIGURE 5.8
Specialist driver set

FIGURE 5.9
Stud remover

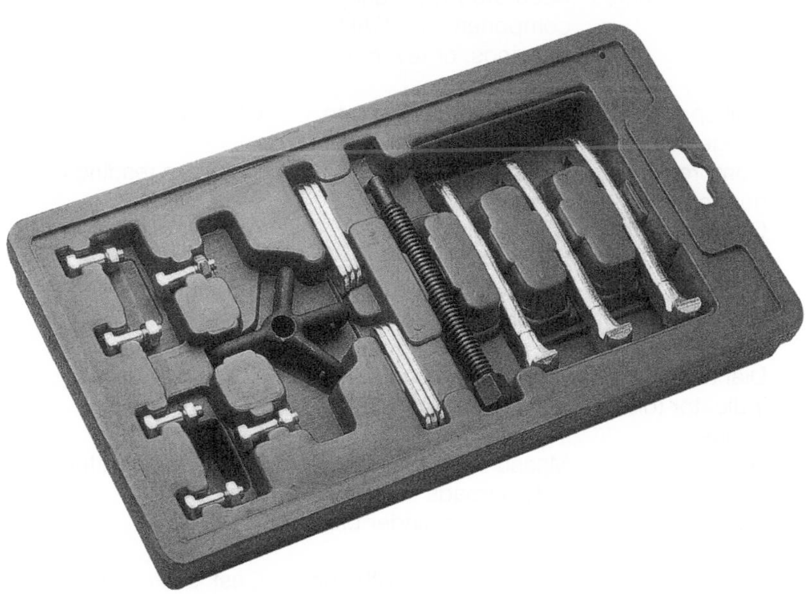

FIGURE 5.10
Hub puller

TABLE 5.3 Overhaul equipment

No.	Item	Purpose	Note
1	4-post hoist – with wheel-free adaptor	Raise vehicle and independently raise one corner	MoT compliant
2	2-post wheel-free hoist	Lift vehicle under chassis so all wheels are free	
3	Body scanner	Scanning bodywork to produce CAD drawings	Good when re-building historic vehicles
4	Cam profiler	Re-grinding camshafts	Make cam profiles to your design
5	Castor, camber and SAI (KPI) gauges	Checking steering geometry	
6	CMM machine	Accurately measuring components for CAD drawings, or reverse engineering	
7	Coil spring gauge	Testing coil spring rate	Check all four springs
8	Compression gauge	Testing engine compression pressure	Compare reading of each cylinder
9	Corner weights	Checking weight distribution at each corner of the vehicle	Adjust suspension and re-distribute weight as needed
10	Crankshaft grinder	Re-grinding crankshafts	
11	Dial test indicator (dial gauge)	Measuring movement – such as for valve lift	
12	Durometer	Measuring the hardness of tyre treads	Check temperature first
13	Engine boring equipment	Re-boring cylinder blocks	
14	Granite table	Providing a smooth and level surface on which to set up the vehicle	Cost is into £1m, used for world-class vehicles
15	Horizontal milling machine	Milling surfaces – such as cylinder head faces	
16	Laser, or light, suspension aligning gauges	Checking steering and suspension alignment	The manual system can achieve the same results
17	Mercer gauge	Measuring the diameter of a cylinder bore	
18	Micrometers – range, internal and external	Measuring inside or outside surfaces – such as cylinder bores and crankshaft bearings	

No.	Item	Purpose	Note
19	Pressure washer	Cleaning mud off the vehicle	Ensure EPA compliance when used
20	Surface grinder	Grinding surfaces such as cylinder head faces	
22	Turntables	Measuring toe-out on turns and carrying out steering checks	Used in conjunction with No. 5
23	Tyre machine	Removing and refitting tyres	Special machine needed with aluminium alloy rims
24	Tyre pressure gauge	Measuring tyre pressures	Need accurate gauge on motorsport vehicles – check temperature before adjusting
25	Tyre temperature gauge	Measuring tyre temperature	Take three measurements on each tyre – inside – middle – outside of tread
26	Vertical milling machine	Milling tasks such as when enlarging inlet ports	
27	Welding equipment (MIG or TIG)	All kinds of joints and repairs	
28	Wheel balancer	Static and dynamic wheel balancing off the vehicle	Use only approved weights – usually stick-on on inside of rim

DATA

Sources of data are very, very important. The obtaining and storage of data is very much a profession in itself. Think of the word library – it does not just refer to books, but all forms of stored information that includes photographs, films, tapes, posters, CDs, DVDs and other electronic storage media.

As a technician in motorsport you will have to collect, collate, store, use and communicate data.

There is a saying that knowledge, another word for data, is power. If that data means 0.1 of a second off each lap, that is very powerful, and therefore very crucial and valuable, data. As a technician, obtaining, using and storing data are high priorities. Table 5.4 looks at the main sources of data.

TABLE 5.4 Data sources

No.	Source	Typical content	Accuracy – limits	Comment
1	Autodata manual	Technical service data	Detailed and checked	
2	Auto Trader	Vehicle sales		
3	Competition vehicle log	Set-up data for vehicle, work done, parts fitted	If correctly maintained	Must be maintained in detail to ensure continuity of operation and save time
4	Customer record/file	Name, address, vehicles owned, contact details	If correctly maintained	Highly confidential
5	Data logger	Vehicle operation data including acceleration, braking, LAG, gear, throttle position	Digital data – as good as system will allow	PI
6	ECU data	Codes for system operation set-up and fault codes	Digital data	Each system may have a separate ECU – a reader will be needed to access data
7	Glass's Guide	Vehicle model guide	Detailed and checked	
8	Haynes (or similar) workshop manual	Vehicle and units service and repair procedures	Detailed and checked	Haynes offer an excellent range of manuals at competitive prices
9	Lap/section time from circuit system	Speed and timing of laps and sections	Digital data – as good as instruments will allow	Data available from race control
10	Manufacturer's workshop Parts Manual	Vehicle and units service and repair procedures	Detailed and checked	Often to be read in conjunction with training materials
11	Parker's Car Price Guide	Vehicle identification and pricing	Detailed and checked	
12	Stack system	Speed and timing of laps and sections	Digital data – made to high level of accuracy	In car timing system – very useful for testing
13	Vehicle service book	Records of service and mileage	If correctly maintained	
14	Your company records	Vehicle changes and modifications	If correctly maintained	Annotated

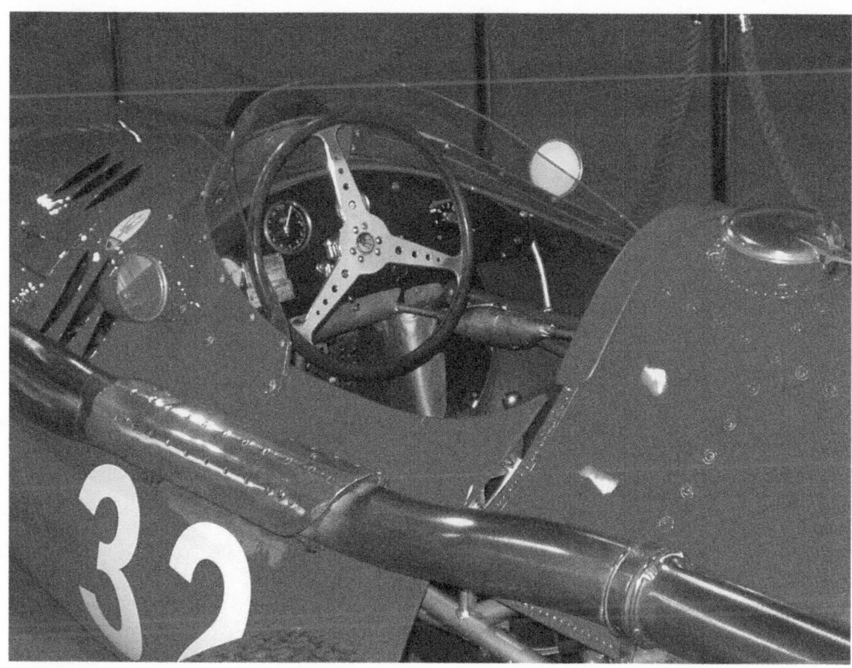

FIGURE 5.11
Maserati 250 – fully overhauled and ready to go

Most manuals and other commercial sources of information are available on CD / DVD, and on-line through a subscription agreement, as well as in paper form.

Before you carry out any overhaul work you will need to check that the data in Table 5.5 is available.

TABLE 5.5 Pre-overhaul data check

No.	Data	Source	Comment
1	Test data	Report on test	This should be added to vehicle log
2	Unit removal data	Workshop manual	
3	Unit test data	Unit manufacturer	
4	Unit stripping data	Unit manufacturer	
5	Equipment operation data	Equipment manufacturer	
6	Replacement parts	Parts manual	Parts used should be recorded
7	Unit assembly and test data	Unit manufacturer	This should be added to vehicle log
8	Set-up data	Vehicle set-up log and/or unit manufacturer	This should be added to vehicle log

FIGURE 5.12
Old and new brake discs off a BMW Mini

PARTS AND FIXINGS

On racing and competition vehicles of all types fixings are very important. It is good when working on a car to remember a bit of engineering science.

1. Force – if you want to fix a post into the ground take a big sledge hammer (14 lb / 6.3 kg mass) and swing it quickly holding the end of the shaft for rapid acceleration; now in terms of a car it is the weight (mass) multiplied by how fast it is accelerating. If you move the formula round you can see that to make it accelerate faster you can either fit a more powerful engine – increasing the force – or make it lighter.
2. Stress – as you can see this is inversely proportional to the area taking the stress – so usually the smaller the part the more stressed it will become.

Bearing these factors in mind – if we make the vehicle lighter and faster we are tending to make the parts more stressed, so we need to be sure that the parts used will cope with the situation. Therefore the parts and fixings tend to be made out of materials which are both light and strong, typically these are high strength steel (HSS), aircraft grade aluminium alloy (7001 or similar series) and titanium. These

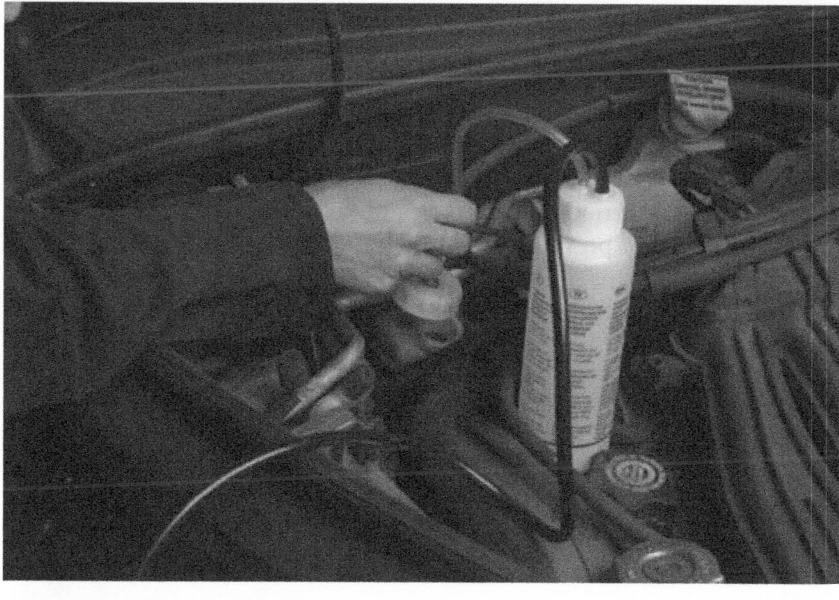

FIGURE 5.13
Brake venting pressure tool

materials are much more expensive than those used in standard cars; also the parts are made in smaller numbers and are therefore much more expensive to produce.

Aluminium alloy and titanium are both easily damaged, and their appearance can soon be marked, so ensure that they are handled with care and that the correct tools are used when working on them.

PROCEDURES

When carrying out any overhaul work you must follow the procedures set out in the appropriate manuals and data sheets. Technical explanations for some tasks are covered in the appropriate chapters.

In all cases follow these basic steps:

- Obtain the necessary data.
- Clean the vehicle in the vicinity of the unit to be removed.
- In a clean work area remove the unit to be overhauled – following the appropriate safety sequences.
- Take the unit to the bench, or mount on a stand.
- Strip the unit, noting the position of parts and using appropriate parts storage trays.
- Repair and rebuild as needed.
- Replace the unit and set-up as per data.
- Thoroughly test and re-check setting against data.

PRODUCTS

The use of products – oils, greases and other chemicals – in the motor industry is big business. The correct use of products can often save time and money for the customer and increase sales revenue for you.

When you have carried out an overhaul task you will probably need some form of product for lubrication or cooling – the customer should be told of this, and if possible given an after care leaflet, or card, with information about the product and its use to prolong the life of the overhauled unit.

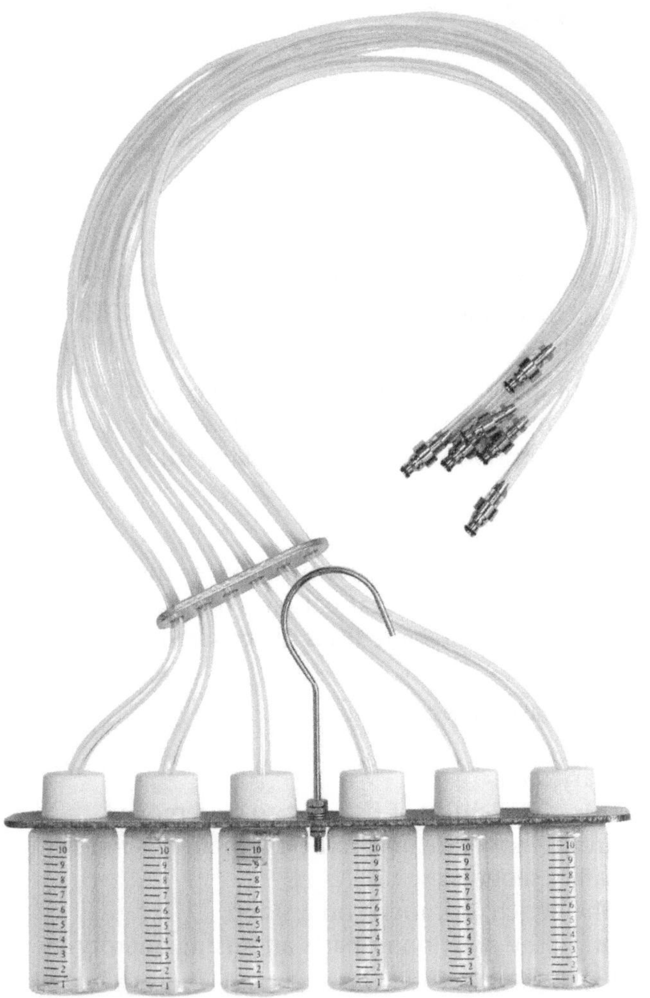
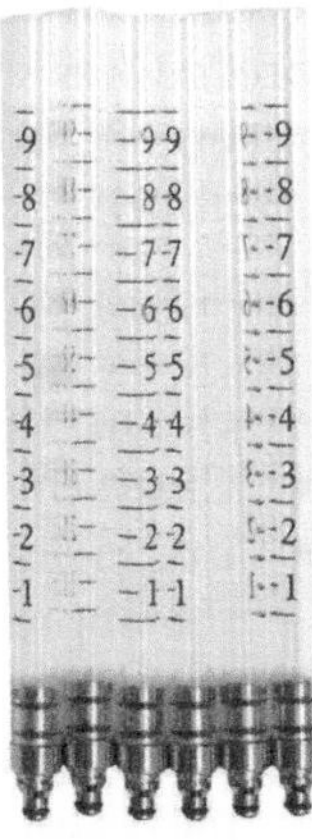

FIGURE 5.14
Fuel injection tester

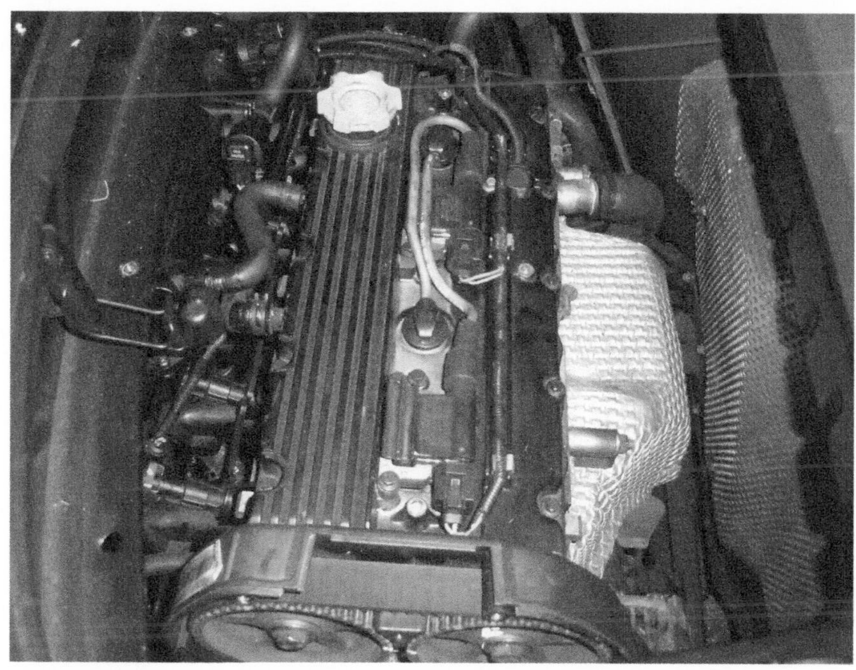

FIGURE 5.15
MGF engine in situ

FIGURE 5.16
MGF cylinder head

FIGURE 5.17
MGF open block

EXAMPLE OF AFTER CARE LEAFLET

BrooklandsGreen – Brake Calliper Care

1. Your brake callipers have been overhauled and should give you a high standard of braking.
2. You are advised to use ***AP Racing DOT 5.1*** brake fluid which complies with SAE J1703 – this is ideal for high performance road, competition and track day use.
3. To prevent corrosion the brake fluid should be **changed frequently** – we recommend at least each racing season.
4. To prevent brake squeal use a **copper based grease** between the pad back plate and the calliper assembly.
5. Be sure to **fully vent** (bleed) your brakes after fitting, before driving.

CLUSTERS – MOTORSPORT VALLEY

The implications of parts supply, knowledge and data, tools and equipment and transportation all revolve around cost. In motorsport there tends to be a team working spirit – yes, even between competing teams. After all, there would be no motorsport if there were not any other teams. So, rather paradoxically, although it is all out on the track, or rally stage, you need to work with your competitors to achieve your aims of winning. Because of this companies tend to work in clusters – small groups based at circuits, in trading estates,

FIGURE 5.18
Close up of wet liners in open block

or sites with good linkages to each other. In the UK the bulk of the major motorsport companies are located in Motorsport Valley. This geographically follows the Thames Valley from Oxfordshire, through the Home Counties, down to Kent. The communication routes use the M4 and M40 leading into the M25 and exiting with the M2.

Networking between the motorsport companies and the competitors – lots of the companies solely exist to be able to race, whilst others operate in other branches of performance engineering, such as aerospace and super yachts to provide an extra income stream – takes place in the Motorsport Valley. Organisations like the Motorsport Industry Association (MIA), the Institute of the Motor Industry (IMI), the Institution of Mechanical Engineers (IMechE), the Motorsport Institute (MI) and *Autosport* magazine organise some of these events. Often they are funded by Government bodies such as UKTI and local development agencies. In the United States of America the magazine and show organiser *Performance Racing Industry* (PRI) is one of the main networking organisations for the industry, they have links with the British Institutions as well as the Society of Automotive Engineers (SAE).

Modifications

Enhance Motorsport System Features

HEALTH & SAFETY AND ENVIRONMENT

Like any other task – you must be aware of Health & Safety and Environmental issues. You should have covered these in detail at Levels 1 and 2. There is a short re-capping in Chapter 5 which is also appropriate to the work in this chapter.

LEGAL IMPLICATIONS

It is important to be aware of the legal implications of any modifications which you make to a vehicle. It is not illegal to sell many parts which when they are fitted to a road going vehicle could be illegal. Also, on competition vehicles, the parts must comply with the requirements of the racing regulations relating to that particular type of racing, or specific class. As a technician, under corporate law, which is vicarious by its nature, if a vehicle to which you have fitted a part is involved in an incident, and the part which you fitted might have been causal towards the damage, then you may be held wholly, or partially liable for the damage caused. That damage could be the death of an innocent person, in which case you could be charged with manslaughter.

So, think carefully about the modifications which you are making. Do not just do it because a customer asked you to do it. Customers are often unaware of the laws, and indeed racing regulations. It is your duty to advise and guide them; you have a duty of care to your customers. Table 6.1 lists some of the commonly broken regulations, or problems.

Some of those points should make you smile – but they are all potentially dangerous and most seriously to be avoided; think the modification through, think safe.

TABLE 6.1 Common mistakes in modifications

No.	Area	Modification	Mistake	Comment
1	Tyres	Racing tyres	Slicks with no tread	Illegal on road
2	Tyres	Asymmetric tyres	Wrong direction of rotation	Unsafe in wet
3	Wheels	Wide wheels	Extending beyond wheel arches	Illegal on road or track
4	Lighting	Fitting spot or fog lamps	Incorrect height or position	Illegal on road
5	Lighting	Any lamps moved by changes to body work	Moved or restricted vision	Check that they comply with position on vehicle and for angle of vision
6	Exterior fillings	Bolts or screws on exterior of body work	Must have a minimum of 2 mm radius and no sharp edges	Common reason for kit cars failing SVA
7	Exhaust	Change system, or parts	Noise limits	Especially appropriate to rally cars
8	Ignition	Change parts	May alter engine operation	Could effect emissions
9	Fuel system	Change parts	May change fuel emissions	Check within limits
10	Brakes	Changing pads	Could alter braking characteristics	Lots of drivers are surprised to find the extra effort needed with competition brake pads
11	Engine	Fit a bigger/ more powerful one	Need to upgrade the brakes and suspension too	I admit to doing this – wondered why it took a long time to stop – very dangerous
12	Suspension	Fit lowered springs	Need to fit shock absorbers to suit	Suspension will bottom
13	Tow bar	Incorrect attachment	Need appropriate fitting kit	Don't forget the electrical connections too

FIGURE 6.1
Head lamp taped for racing

The next sections looks at frequently modified areas of the vehicle – as specified in the IMI Criteria – discussing typical modifications and the reasons for these modifications

REASONS FOR MODIFICATIONS

The reasons why customers modify, or enhance, their vehicles are many and varied; but usually they can be classified as one of the following for making the vehicle:

- faster
- more powerful
- lighter
- handle better
- stop more quickly
- more comfortable to drive – especially under specific conditions, such as extra lights for night driving
- look good – attractive
- comply with specific racing regulations.

A good motorsport technician will know all the right tweaks to enable the car to perform better than the other competitors.

FIGURE 6.2
MGF Bonneville record breaking car

CYLINDER HEAD

Cylinder heads may be made from cast iron (CI) or aluminium alloy. The aluminium alloy ones are about a third of the mass (weight) of the equivalent CI ones. Aluminium is a better conductor of heat than CI and therefore more suitable for high-performance engines which produce more heat for the equivalent capacity. Remember that when petrol and air, or diesel and air, is burnt it produces heat. About 40% of the heat does useful work in pushing the piston down the bore; 30% goes to the coolant jacket and 30% to the exhaust. If you increase the power output of the engine then you are also going to increase the heat to the coolant and the exhaust. Changing to an aluminium cylinder head will improve the rate of heat conduction and so help to prevent engine failure through overheating. If you are changing a cylinder head for this reason then it is also worth checking and if necessary changing the:

- hoses – high pressure
- hose clips – high clamping pressure
- coolant – inhibitor for use with aluminium
- use of wetting agent
- radiator flow capacity
- coolant (water) pump – flow rate

- coolant pump drive belt and velocity ratio
- cylinder head gasket – check material, thickness and use of rings
- cylinder head studs or bolts – need high tensile strength ones.

Both types of cylinder heads may be modified to accommodate larger valves by machining; the valve seat angles changed, the compression ratio increased by skimming; and the ports smoothed and polished to improve gas flow.

VALVES

The purpose of the valve is to open and close to control gas flow. It's opening and closing speed is limited by its mass, remember Newton's Second Law, F = Ma. So the valve needs to be as light in weight as possible – for this reason titanium is frequently used. The higher the engine output the higher the amount of heat which is generated. So the valves need to be able to dissipate heat – for this reason sodium filled valves are sometimes used as the latent heat of the sodium filling serves to reduce the valve temperature.

Valve shape is also of consideration, in terms of valve seats and head shape to improve the flow of the gasses both into and out of the cylinder head.

FIGURE 6.3
Hot rod made from a bomber plane fuel tank

VALVE SPRINGS

The valve springs close the valves. The camshaft opens the valves, so the faster the engine revs, the faster they will open. Remember that if we can make the engine rev faster it will probably be able to develop more power – within certain restrictions. So we are now dependent on the springs closing the valves – if we are increasing the engine speed we will need to fit stronger valve springs. Otherwise the springs will not close the valves fast enough and the engine will suffer from valve bounce – indeed the valves may touch the piston crown. To modify the valve spring closing rate there are usually three options – stronger single springs, double springs, or if double springs are already fitted, triple springs can be fitted.

Fitting double or triple springs may only be possible if the valve retaining caps are changed to match and new springs and the spring seat on the cylinder head may need machining to accommodate the new springs.

CAMSHAFTS

The camshaft controls the valve opening in terms of both lift and period. That is:

- Valve lift – the distance the valve is lifted from it's seat. Like opening a door – the greater the lift the more gas that can be put through at any one point of time. The valve lift may be the same as the lift at the cam, or there may be a ratio so that the valve lift is greater than the cam lift – when using a rocker mechanism. Direct acting cams do not have this option.
- Period (also called duration) is the number of degrees that the valve is open – measured between the opening and the closing points. The number of degrees will correspond to a period of time for any given engine speed. Again – like a door – the longer it is open the more petrol and air that can get through it. If the valve period is increased from say 100 degrees to 110 degrees it will be open 10% longer, so 10% more gas can pass through it. This is often simply referred to as valve timing or just timing.

There are three main options:

1. Change the camshaft for one with different opening periods and/or lift.
2. Change the rocker mechanism ratio.
3. Change the cylinder head and camshaft for one giving a different location – for instance fitting a double overhead cam (DOHC) cylinder head in place of an overhead valve (OHV) arrangement.

Also to be considered is the material from which the camshaft is made. Many popular vehicles use cast iron because it is easily made, cheap and sufficiently hard for normal usage. For competition use it is normal to use a high grade of steel with sufficiently high carbon content so that it can be induction hardened.

CAMSHAFT DRIVE

Camshaft drive is usually a toothed belt or chain, with a small number of vehicles using gear drive. Table 6.2 compares the different types.

TABLE 6.2

No.	Type	Advantages	Disadvantages	Comment
1	Toothed belt	Simple, no lubrication needed, adjustable, quiet in operation	May break unexpectedly	Most popular
2	Chain	Long lasting, adjustable, wear clearly visible	Needs constant lubrication, expensive	
3	Gear (metal to metal)	long life	Noisy, expensive, needs constant lubrication	
4	Gear (metal to fibre)	Quieter than (3) and does not need lubrication	Fibre gears needs frequent replacement	Used on racing motorcycle engines

CYLINDER BLOCK

Cylinder blocks are usually the subject of detailed regulation for most competition classes – because the block is usually the largest mechanical component and limits the power through its capacity (swept volume), configuration and general structure.

Material – aluminium alloy is the best choice for low weight and thermal conductivity. This may be either die-cast or sand-cast. The grain structure of aluminium alloy engines can be improved by head treatment – this can include deep freezing for several days.

Capacity – changing the bore and/or the stroke will alter the engine capacity. Usually there are competition limits on over boring, and over boring will weaken the structure of the engine.

Configuration – that is number and layout of cylinders. Now before you say that not much can be done about this point, it must be remembered that the first Cosworth V8 was made from two four-cylinder engines put together, and SAAB made a range of engines using the same components in different configurations.

Liners – these may be of the wet type or the dry type. The liners may be coated or plated to reduce friction, or increase life. One common material is chrome – its shiny surface reduces friction and therefore can increase power output. The liners may also be made out of different materials, steel is normal for motorsport vehicles. If a cast iron cylinder block is being over bored – taken to a size above normal re-bore limits – then it is often good practice to machine it out even more, then to add steel liners. This gives better durability and ensures a minimum cylinder wall thickness.

When new dry liners are fitted into a block, the fitting is usually an interference fit, that is, a press fit. This may cause irregularities on the inside of the cylinder – these are removed by honing with a very fine grade oil stone. Of course, this procedure cannot be carried out with coated liners.

Deck height – the height of the top of the cylinder block in relation to the piston crown. The deck height affects the compression ratio. A minimum deck height of about 0.2 mm (0.010 inch) (always check manufacturer's figures) is needed to prevent the piston crown touching the cylinder head due to the dynamic forces at maximum engine speed.

Sealing – increasing the engine's power output will put a greater pressure load on the engine's seals. Some engines are built without

FIGURE 6.4
Mini grinding stones

gaskets, as gaskets are used to allow for uneven surfaces these engines are machined very accurately. A sealant is used between the mating surfaces.

Cylinder head gaskets may be replaced on motorsport engines by ones made from different materials – frequently this is one sheet of malleable metal. Where the compression ratios are very high the block may be machined with annular rings around the cylinder bores to accommodate sealing rings.

The sump and other gaskets may be made from materials which are more able to cope with the high pressures and temperatures of competition engines – they will also be made to more exacting levels of accuracy than standard road versions.

Bearings – the crankshaft main and big end journals are likely to be harder than those of road cars, so the bearing surfaces must be able to work with the materials. Normally the bearings are harder, containing more tin and less lead to make up the white metal alloy.

The bearing caps may also be upgraded by either fitting steel caps which are then line bored to match the crankshaft, or fitting steel straps over the existing caps to increase their strength.

PISTONS

For high performance purposes the piston is a very important part of the modification process. Table 6.3 sets out the characteristics of piston design.

TABLE 6.3 Characteristics of Piston Design

No.	Part of piston	Design characteristic	Reason	Comment
1	Crown	Height above gudgeon pin	Alter compression ratio and relation to deck height	
2	Crown	Shape	Changes to swirl and cut outs for valve pockets	Valve pockets may need to be machined to cope with extra valve lift
3	Gudgeon pin	Diameter	To cope with extra load	This will also involve modifications to the con rod

(Continued)

TABLE 6.3 ***(Continued)***

No.	Part of piston	Design characteristic	Reason	Comment
4	Gudgeon pin	Type of fit	To suit con rod	Types of fit include fully floating and interference
5	Skirt	Shape	Slipper skirt to reduce friction with cylinder wall and reduce weight	
6	Skirt	Design	Solid skirt to increase strength	
7	Skirt	Length	Short skirt to reduce friction with cylinder wall and reduce weight	
8	Rings	Number	Less rings – two or three – to reduce friction with cylinder wall	
9	Rings	Type	To enhance combustion sealing	Ring material may also be changed
10	Material	May be forged, billet or cast from a variety of aluminium alloys	Will effect weight, strength and machining processes	Pistons can be made any size or design; but if non-standard will cost accordingly

As an overview, there are only a small number of companies that produce pistons in the UK and America, and not withstanding the Asian market, they tend to produce for the world-wide motorsport range too. This is because piston production is very specialised precision engineering requiring expensive machine tools and highly skilled staff. The motorsport piston supply is largely related to modification of a standard piston – although this is done at the manufacturing stage, not as a post-production change. Or, the application of a piston to one engine that is used as standard in another engine. If you think of the Ford range of engines that come in various capacities and are used in different models you may get a better picture.

CONNECTING RODS

The most common modification to connecting rods (con rods) is balancing them. There are a number of different ways of balancing con rods. The basic principle is to ensure that they all weigh the same amount, and that the weight of the top (little end) and the bottom (big end) is also equal on each con rod. This is done by supporting one end and weighing the other. To lighten the con rod, metal is moved by drilling holes in lesser stressed sections. Weight can be added by drilling holes and filling them with lead which is much denser.

The con rods offer resistance to the engine, both from their mass – resistance to accelerate – and the air resistance moving inside the crankcase. The acceleration is improved by using lighter materials; in ascending order of cost the options are: aluminium alloy, titanium and carbon fibre. The aerodynamics is improved by making them an aerodynamic shape.

An economical improvement can be made to the original con rods by balancing, shot blasting smooth then polishing.

If the stroke length is being changed by changing the crankshaft – a very popular modification on many engines, then it may be necessary to change the con rod to ensure that the piston operates in the correct area of the cylinder bore.

When building engines, no matter which type of con rod is being used – you should check:

- the oil way drilling
- the little end fastening
- for straightness
- the big end fitting, especially the cap and retaining nuts and stud or bolts.

More big end caps come loose than con rods break in the middle. Use the correct fastenings and torque to the correct setting. As each piston and con rod is added to the block check that the engine turns freely. Also ensure that the little end fit is correct – as if it is not, the piston is likely to break at this point. That is the piston boss breaks away from the piston skirt and the engine is likely to be wrecked.

On very high-performance engines the con rods should be seen as a service item – that is replaced at each rebuild along with the bearings.

CRANKSHAFT

Materials – the crankshaft is one of the main components in limiting the engine speed and power output. Standard crankshafts may be made from either cast iron (CI) or steel. The steel is likely to be a forged medium carbon variety. For high performance applications the steel is usually upgraded to one which will accept surface hardening readily. This usually means a slight increase in carbon content and the addition of other metals such as chromium, silicone, copper and aluminium. The two main types of surface hardening, to increase crankshaft life by reducing wear, are:

- Nitriding – that is immersing in a *salt bath* – that is a solution of nitric acid (very hazardous) for a controlled (long) time period at a pre-set temperature. This is usually applied to more expensive steel alloys and increases the strength of the crankshaft.
- Tufftririding – a cheaper option for use with cheaper steels and only giving a surface hardness. This involves immersion of the crankshaft into a bath of (very hazardous) sodium cyanate at 570 degrees Celsius for two hours.

Design – the design of the crankshaft has a number of factors for consideration when modifying the engine, the main ones being:

- Configuration – this affects the firing order and balance, on six and eight cylinder engines there are a number of options.
- Webs and counter balance – for high speed engines these are removed to make the crankshaft light, balance becomes dynamically controlled.
- Balance – done on a balancing machine both statically and dynamically, usually in conjunction with the flywheel, con rods and pistons.
- Thrust races – to prevent longitudinal movement, thin needle roller bearings may replace the plain metal ones to reduce friction.
- Oil ways – the oil passage ways are cross-drilled, that is extra holes are drilled across the crankshaft to ensure a good oil supply to the big end bearings.

FLYWHEEL

As the purpose of the flywheel is to keep the engine turning between firing strokes, the faster the engine runs the less the flywheel inertia is needed. So faster running engines, and ones with more cylinders require flywheels with less mass (weight). The less the mass of the flywheel the faster the engine will be able to accelerate. However, this will lead to uneven running at low engine speeds – not a problem on race cars.

The flywheel also locates the starter ring gear and provides one surface of the clutch assembly.

FIGURE 6.5
Mini Cooper S modified dashboard

Modified road cars – the process is to maintain the original flywheel lightening it by removing surplus material from less stressed areas like the back of the flywheel towards the outer edge.

Full race cars – the flywheel can be replaced by a thin sheet of plannished plate (usually hammered or rolled to be stiff and flat) with a flanged edge to accept the starter ring gear, and a small diameter multi-plate clutch pack is used which does not need the flywheel as a surface for friction.

Balance – the flywheel will be balanced both statically and dynamically in conjunction with the crankshaft and other rotating parts.

Be aware – changes to the flywheel may have an effect on dynamic balance and it may also be necessary to look to the crankshaft front pulley, TV damper and any counter balance shaft mechanism.

LUBRICATION

The first modification to the lubrication system is to use a good quality oil and oil filter. There are lots of alternatives for each type of engine, and lots of different makes of each type. The supply of oil to the motorsport industry is both high value and high profit. So some detailed care in the choice of oil is needed. Many manufacturers have

TABLE 6.4 Examples of API Oil Classification

No.	API Classification	Application	Comment
1	SN	Cars made in 2011 or older	Check for updates
2	SM	Cars made in 2010 or older	
3	SL	Cars made in 2004 or older	
4	SJ	Cars made in 2001 or older	

a preferred brand which is often one of their sponsors. For example Ferrari recommend Shell and Shell sponsor Ferrari. Two of the major makers are Castrol and Duckhams with Comma coming up the grid.

You will remember from Level 2 that oils are classified by the Society of Automotive Engineers (SAE an institution based in America) for viscosity (how they flow) and that the common grades are 10W–40 and 15W–40. The capability of a particular make of oil to do its job in an engine is classified by the American Petroleum Institute (API), Table 6.4 gives some examples, but see their website for more details (useful websites are listed on page 195).

Actual mechanical modifications which are typically made to the lubrication system are:

Relief valve – the operating pressure of the relief valve may be changed, this may be by changing the spring, adjusting the spring setting or replacing the complete assembly. Usually increasing the oil pressure by about 10% will cope with greater bearing loads caused by engine power increases. Typical oil pressure is 60 to 100 psi (4 to 7 bar).

Oil cooler – this is to control the oil temperature and prevent engine damage. Typical oil temperatures in degrees Celsius are: road car 100–120; high performance car 120–150; race car 150 plus. The oil cooler must be fitted in a place where cold air flows freely – often this is just in the form of the engine coolant (water) radiator. The flow of oil through the cooler may be controlled by a thermostat. The oil flow is usually in series with the oil filter.

Sump baffles – these prevent oil surge on corners, so preventing oil starvation because the oil has moved away from the oil pick-up; or preventing oil burning by it climbing the cylinder walls. Baffle plates may be welded inside an ordinary sump.

Oil pick-up – the modifications to this part include fitting a broader mesh filter to allow greater oil flow and moving the open end to the centre of the sump to avoid oil starvation (see sump baffles).

Oil filter – a high flow filter may be fitted to improve oil flow or it may be remotely mounted for ease of changing and incorporation with the oil cooler.

Dry sump – instead of the lubricating oil being in the sump underneath the crankcase a separate oil tank is used. The oil from the engine components fall into the sump; but it is drawn out of the sump to the tank by a separate – scavenge – pump. The pressure pump draws oil from the tank to pressure feed the engine's bearings. This gives the advantages of: lower centre of gravity as the engine can be lowered in the chassis as the sump is very shallow; extra cooling of the oil; reduction of surge and foaming problems.

CARBURETTERS

No new cars have been made with carburetters since the end of the 1990s because of the vehicle emission regulations. However, many old competition vehicles have carburetters and some special builders use them on track-day and race vehicles because they are cheap, easy to understand and can be adjusted manually with the minimum equipment in the paddock.

So, although not part of the syllabus, a motorsport technician could be expected to know a little of the background of the common motorsport carburetters.

SU – the oddly named Skinner's Union. This variable choke, constant depression carburetter is a very popular choice for modifying engines.

Stromberg later made a similar carburetter called the CD (constant depression).

Weber – designed by Edoardo Weber who first made carburetters for the racing FIATs of the 1920s, then they were fitted to the first Enzo Ferrari built car bearing his own name in 1948. The Weber carburetter's unique design point is that it is a fixed choke carburetter body on which every single part can be changed. **Dellorto** introduced a similar carburetter in the 1960s.

Zenith – a fixed choke carburetter made in a number of different styles, very popular on many cars and eventually combined with the similar operating **Solex**.

Holley – originally made in Manchester, England but moved to America and now found on most American V8 engined race cars.

Amal – originally designed for motorcycle use; but used on many small engined race cars such as the Mini. Its unique design point is the integration of the choke body concentric with the float chamber.

Minnow-Fish – a precursor of the fuel injection throttle body using the low pressure 3 to 5 psi (0.2 to 0.3 bar) fuel supply from a normal petrol lift pump.

Main modifications – the main modifications of carburetter arrangements are:

- increasing the number of carburetters – ideally one per cylinder
- increasing the size of the carburetters
- changing the air filter (cleaner) for a high flow one – for example K&N
- fitting air trumpets of different lengths
- fitting water heated, or non-heated inlet manifolds
- altering the carburetter setting to give different mixture strengths under different running conditions.

INJECTORS

Fuel injection systems are typically modified in two ways:

- The Electronically Programmable Read Only Memory (EPROM) in the fuelling ECU is re-programmed to give more fuel under particular conditions. This is called *chipping* and the EPROM is a type of micro-chip. Chips can readily be bought for most popular performance cars. On rally cars which have off-road performance requirements separate to their on-road legal requirements it is often the case to have two separate, but switchable ECUs.
- To deliver the extra fuel, high flow rate injectors are used when the engine is chipped. These are sometimes identified by the colour code green as against the normal grey.

THROTTLE BODIES

Instead of the normal injectors and manifold arrangement, each cylinder is fitted with short individual stubs which incorporate the injectors. This improves air and petrol flow.

INLET MANIFOLD

These may be changed for ones which give different air flow characteristics and carburetter or injector arrangements.

AIR FILTER

The air filter (also called air cleaner) must be appropriate for the vehicle and its operating conditions. Remember from Level 2 that it removes dirt from the incoming air to reduce the risk of engine damage, quietens the noise of the air intake and acts as a flame trap.

For track use an air filter may be thought unnecessary and intake trumpets used instead. For a road car a high flow system such as the popular K&N system may be used. For an off-road vehicle a high level – to keep out of the water when wading – system may be used with two air filters in series to reduce the risk of an ingress (taking in) of dirt or dust. One of the air cleaners may be a centrifugal one.

If twin turbo-chargers are used then two air filters may be appropriate.

EXHAUST SYSTEM

The exhaust system has the job of quietening the noise of the exhaust gas and cooling it before it leaves the vehicle. It will also, except on older vehicles, incorporate a catalytic converter (cat) and a lambda sensor. Most modifications relate to improving the flow of the exhaust gas from the engine to the outside of the vehicle. Some are solely about altering the exhaust note to make it sound *sporty*. The exhaust note is very important to some people – about impressions of power. It is worth mentioning that Alfa Romeo has a special department to ensure that their cars all have the correct corporate image exhaust note. It is good to listen to different exhausts and see if you can identify the cars by them.

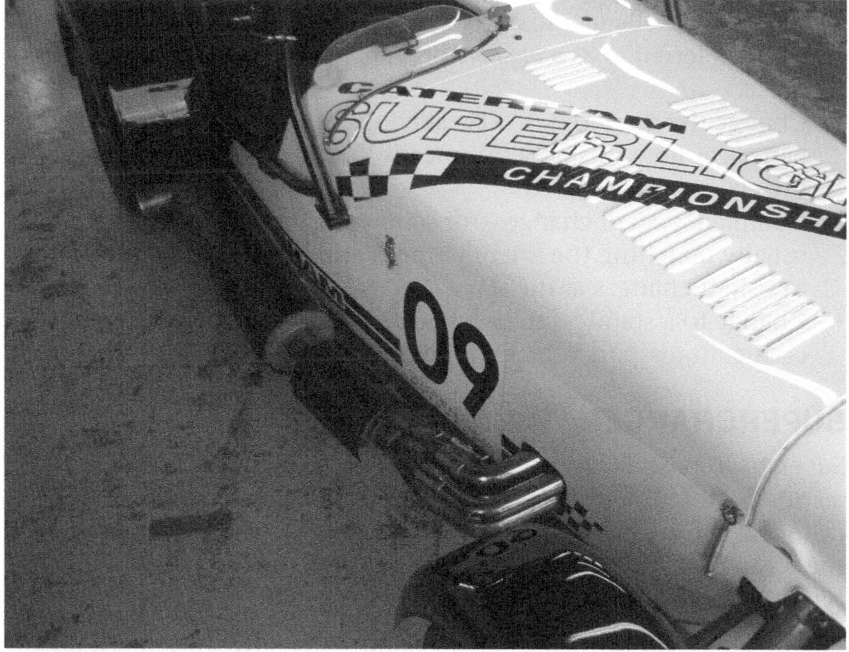

FIGURE 6.6
Four into one exhaust manifold

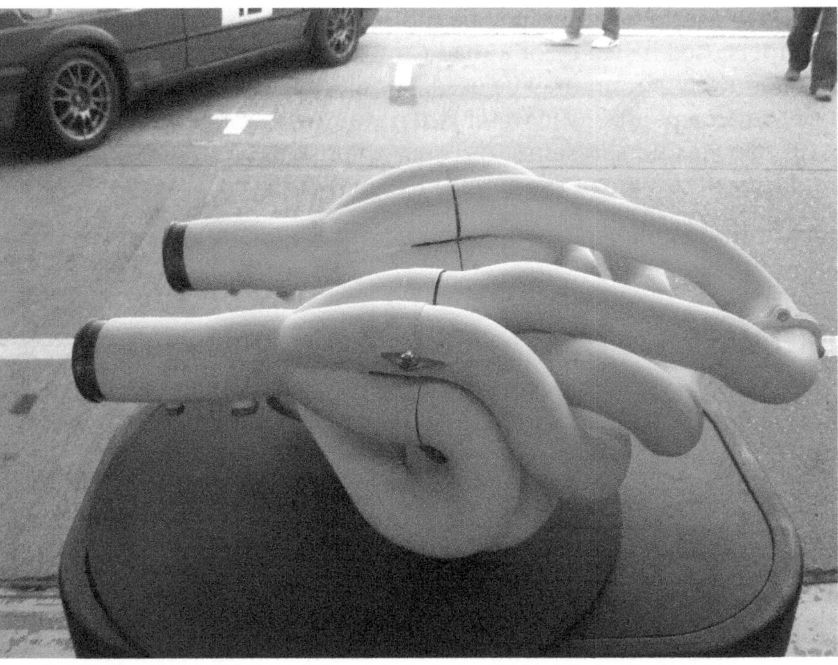

FIGURE 6.7
Coated exhaust manifold

Typical exhaust modifications include:

- Change rear silencer (back box) for one with a larger diameter tail pipe – referred to as a *big bore* or *drain pipe*. This is for looks and noise.
- Change the manifold (in US called headers) for one to give a better gas flow – usually this means longer individual pipes before the cylinders are joined.
- Complete system change – for serious performance increases, usually matching the inlet manifold and cylinder head and camshaft changes at the same time.
- Change to a stainless steel exhaust system – most appropriate to classic motorsport vehicles for longevity.

SUPERCHARGER/TURBOCHARGER

Superchargers and turbochargers are fitted to increase the amount of air and petrol (air only on diesel engines) going into the engine cylinder. The greater the amount of gas that goes into the cylinder the greater the power output will be. Superchargers and turbochargers typically increase engine output by 30%.

The supercharger is in effect an air pump driven by the crankshaft – they started life on aircraft where the air is less dense and needed to be compressed both for the engine and the passengers.

The turbocharger is driven by the exhaust gas, it has two turbines. One turbine is driven by the exhaust gas; the other turbine is driven by the shaft from the first turbine and compresses the air into the cylinder. On high performance, and diesel vehicles, an intercooler may be used to cool the air between the turbocharger and the cylinders.

Although superchargers and turbochargers do the same job, they do it in a very different ways. Table 6.5 overleaf compares superchargers and turbochargers.

WHEELS AND TYRES

The choice of wheels and tyres is almost endless. The current trend is to go for large diameter wheels with very low profile tyres. The resultant small side walls allow high speed ratings. When fitting replacement wheels and tyres you must check the following points to avoid failure and possible damage:

- Confirm the wheel and tyre manufacturer's fitting is approved to the vehicle concerned.
- Check the off-set for chassis and body work clearance.
- Check hub/flange/stud fitting and bearing load.
- Check the speed rating.
- Set the pressures accurately.
- Torque up correctly using the designated nuts and studs.

STEERING

Rack – the normal modification is to fit a quick-rack, which is one which requires less turns from lock to lock than the standard one.

Geometry – this may be modified in a number of ways: changes to the castor, camber, SAI (kpi) and toe-in. On some vehicles there is a facility for adjustment. On track cars the use of adjustable rose joint is the usual method. To improve handling the use of negative camber is common.

BRAKES

The braking system converts kinetic energy into heat. The limiting factor with braking systems is the maximum rate of heat dissipation – how fast they can conduct and radiate the heat away. On top flight race and high-performance vehicles the braking components are made largely from carbon. This is not, however, suitable nor affordable for most motorsport vehicles. Typical modifications of the main components are:

Discs – replace with ones which are ventilated, cross-drilled, or larger diameter. In all cases the aim is better cooling.

TABLE 6.5 Comparison of supercharger and turbocharger

No.	Type	Drive	Characteristics	Advantages	Disadvantages	Comment
1	Supercharger	Belt from crankshaft	Straight line pressure increase dependent on engine speed	Immediate response to throttle	Needs drive from crankshaft	Ideal for rapid acceleration – dragsters
2	Turbocharger	Exhaust gas	Delayed pressure increase dependent on exhaust gas pressure and A/R ratio	Exhaust drive giving better economy for given application	Delayed response to throttle – prompts use of twin turbochargers	Ideal for application combining power and economy – rally cars and diesels

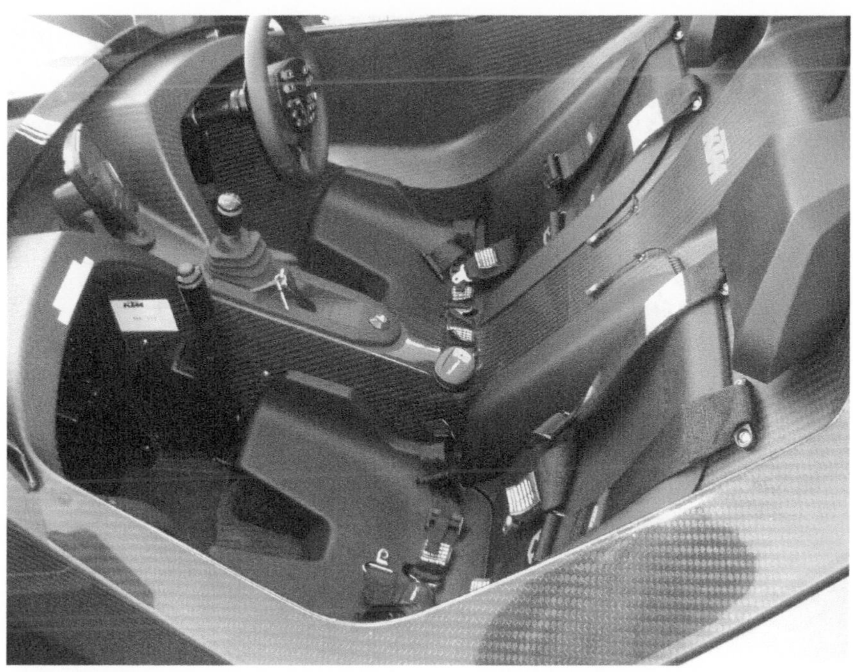

FIGURE 6.8
Plastic KTM seat in carbon fibre tub

Callipers – replace with ones which have more pistons to increase the applied pressure.

Hoses – typically add braided flexible hoses and stainless steel lines to give less hoop (that is outward like a hoop) expansion when maximum pressure is applied.

Fluid – use a high boiling point fluid to prevent brake fade. Remember to test brake fluid for water content – hygroscopic test.

BODY

Competition cars are likely to have bodywork modifications for a number of different reasons. Common reasons are:

- light weight – use of GRP, composite and plastic components
- suspension changes – addition of new mounting points
- engine/gearbox changes – new or additional mountings
- wider wheels – extended wheel arches.

SAFETY DEVICES

The modification is likely to be the addition of a roll-over bar and fire extinguisher system. Although for short circuit cars the use of window nets and front mesh screens may be appropriate.

FIGURE 6.9
Fire extinguisher system

INTERCOM

Rally cars are likely to have driver-to-navigator intercoms.

LOG BOOK

Changes to the vehicle should be recorded in the vehicle log.

CLUTCH

The clutch transmits the power from the engine to the gearbox – it must be capable of doing this. Typical clutch modifications include:

Operating mechanism – the operating mechanism may be changed – an hydraulic linkage can be set up to provide a better mechanical advantage as well as cope with different clutch positions to that of a cable mechanism.

On motorcycle engine systems, such as Suzuki Hybusa and Yamaha R1 engine and gearbox set-ups, the need to disengage the clutch is removed on up changes by having a system which cuts the fuel supply when the gear change is moved. This slows the engine momentarily to allow a smooth clutch-less change, saving effort and time.

Clutch springs – to transmit the extra torque of a tuned engine the clutch springs will require a higher force when using the same size and type of clutch plate.

Clutch spinner plate – for competition purposes a solid spinner plate will be used. This may be of the paddle type for reduced rotating mass – hence quicker gear changes as the speed of rotation can be changed more quickly.

Clutch material – a variety of materials are used for clutch friction plates (spinner plates), these include carbon, and ceramic – often sintered to the steel and bronze.

Multi-plate clutch – a multi-plate clutch mechanism allows the transmission of greater torque by the use of more friction surfaces. This also allows the construction to be of smaller diameter – contributing to a reduction of rotating mass and hence better acceleration. Multi-plate clutch kits are available for most engines commonly used in motorsport.

GEARBOX

The main change with the gearbox is to get the correct ratios. For a circuit car it is common to have different gear sets available for each different circuit. This is easily changed with a Hewland type gearbox where the gear cluster is removed in one unit assembly from the rear of the gear box without removal from the car.

On road cars, often the only alternative is an after market dog box – these are usually simply close-ratio gear sets to enable the use of a narrow power band.

DRIVE SHAFTS

Up-rated drive shafts and joints may be needed if the power output is increased. With popular models of vehicles it is often the case that the drive shafts, joints and the suspension and brake assemblies from the more powerful models can be used on the tuned lower range models.

ENGINE MANAGEMENT

ECU – the main modification in this area is chipping (see injectors). As well as modifying the fuelling chip the ignition chip may also be changed to alter the burn time, ignition timing and virtual dwell. This activity is also called mapping as the 3D representation resembles the contours of a map.

Spark plugs – the main change is the use of harder, or colder operating, plugs to prevent the engine from overheating and the onset of pre-ignition.

Always ensure that the spark plugs have the correct reach, the correct sealing arrangements, the correct centre electrode assembly, and are installed to the correct torque setting.

Sensors – more accurate sensors may be used, or ones which operate in a different range.

Fuelling – see injectors.

POWER SUPPLY

Alternator – rally cars and off-road vehicles usually need larger output alternators to power the additional electrical equipment.

Battery – rally and off-road vehicles are usually fitted with larger capacity batteries to provide a bigger reserve of electrical power for the additional electrical equipment. Often the electrical equipment will be draining the battery when the engine is not running, or running slowly so that the charging rate is below the drain of the electrical equipment.

Where motorsport vehicles are likely to roll, that means most of them, then the use of gel batteries is recommended so that there will not be a loss of electrolyte. Also ensure that the battery box complies with the appropriate competition regulations – normally this means some sort of fully enclosed box with a breather to the outside of the vehicle.

AUXILIARY SYSTEMS

Motorsport vehicles may be fitted with a variety of auxiliary systems, typically data logging, navigational aids and communication systems. When fitting these ensure that the instructions are followed in detail and that any legal requirements are fully met.

LIGHTING

Additional lighting when fitted must comply with legal and competition regulation requirements.

SHOWS AND OTHER SOURCES

Gaining detailed information on motorsport vehicle system enhancements requires some individual research – especially if you are interested in one specific make or type of vehicle. You will need to visit shows and websites.

Shows – the major shows to visit are:

- *Autosport International* – held at the NEC Birmingham every January, includes F1 and all categories of clubman racing too.

FIGURE 6.10
Painted on headlamp on a Toyota racing car

- *RaceRetro* – held at Stoneleigh, Warwickshire in early spring, covers historic racing cars and motor cycles – these are the most contested classes of racing in both the UK and the USA.
- *London Motorshow* – held in July at the Excel Centre, London.
- Motorcycles shows – there are four main ones: Alexandra Palace, London; Excel Centre, London; NEC, Birmingham and probably the biggest, but out doors, is *Motor Cycle World* held at the National Motor Museum, Beaulieu, Hants in July.

Websites – all the performance car and motorcycle component manufacturers and many other related companies, have websites. You will find suggested websites listed on page 195.

CHAPTER 7

Electrical and Electronic Principles

This Chapter covers the motorsport electrical and electronic principles for both the Engine and the Chassis units

The electrical and electronic systems on motorsport vehicles fall into a number of different and overlapping classifications, these are shown in Table 7.1 overleaf.

Safety – working with electrical and electronic systems is not easy for many motorsport engineers, the reason being that you cannot see electricity except when it sparks – and often that is too late, the

FIGURE 7.1
Paulio Racing's handmade dashboard

TABLE 7.1 Vehicle System Identification

No	Type of vehicle	Basic lighting	Advanced lighting	Engine electronics	Chassis electronics	Transmission electronics	OBD	Data logging
1	F1		X	X	X	X	X	X
2	Historic	X						X
3	Le Mans		X	X	X	X	X	X
4	Motorcycle – trial	X		X				
5	Motorcycle – circuit			X	X		X	X
6	Rally car – club		X	X	X		X	X
7	Rally car – WRC		X	X	X	X	X	X
8	Single seat	X		X			X	X
9	Sports		X	X	X	X	X	X
10	Touring car		X	X	X	X	X	X
11	Trial car	X		X			X	X

damage is done. The following points should guide you safely when working on any vehicle:

- If any fault is suspected, the first job is to use the on-board diagnostics. If this shows a problem than you have the answer.
- Before you start to disconnect any wiring, or component, disconnect the battery (often this can be done by turning the isolator switch), then wait for two minutes for any stored capacitance to discharge. If you are working on a SRS (air bags) then you must be very careful as you need to wait for two minutes for the current to leave the circuit or the air bag may discharge. Treat the anti-submarine seat belt pre-tensioner in the same way.
- Data logging systems should be downloaded onto your laptop before you carry out any work on the electrical system for fear of loosing the information in them – rapid discharges of electricity or other electro-magnetic interference can cause a loss of stored data.
- Do not use a mobile telephone near a memory device – this too can destroy the data.
- Keep HT (ignition) devices separate from LT and memory devices to prevent interference.

FIGURE 7.2
JCB diesel world land speed record car dashboard

GENERAL PRINCIPLES

In this section we will re-cap on some of the electrical terms and formulae.

Volt (V) is the measure of electrical force (E); potential difference (PD) is measured in volts as is volt-drop and electromotive force (EMF). Volt drop allows us to calculate the resistance of wiring of an electrical component.

The nominal voltage of most vehicle circuits is 12V, if the battery is fully charged it will read 14.2V, less than 12.6V and it is discharged and less that 10.5V it probably has a fault. ECUs tend to run at 5V, dash instruments at 9V or 10V and some electronic components at 2V. HT spark is 5kV to 10kV on single-coil systems, 40kV or more on systems using ballast resistors, electronic boosting or multi-coils.

Ampere (A) is the measurement of the quantity of electricity, called current (I). Batteries are rated in the number of Amps they can provide over a time period (see battery). Ignition systems use about 0.5A, lamps about 5A each, the starter motor needs about 170A to turn over the engine from cold.

Resistance (R) is measured in Ohms; it is the slowing down of the flow of electricity. Ohm's Law states that current flow (I) is equal to the electrical energy (E) divided by the resistance (R)

$$I = E/R \text{ (also written } I = V/R)$$

$$\text{So that: } V = RI$$

Electrical power (P) is given in Watts (W), or kW (1,000W)

$$P = EI \text{ (also written } P = IV)$$

$$\text{Watts} = \text{Volts} \times \text{Amps}$$

As

$$V = RI$$

$$P = I(RI)$$

So it can also be written

$$P = I^2R$$

In **series circuits** the resistance is additive:

$$\text{Total Resistance} = R1 + R2 \ldots\ldots\ldots\ldots.$$

The voltage across each resistor is proportional to the resistance, using Ohm's Law. The sum of the voltages across each resistor is the total circuit voltage.

Kirchhoff's voltage law

$$\text{Total Voltage} = V1 + V2 \ldots\ldots\ldots\ldots.$$

If one resistor in a series circuit fails the entire circuit becomes dead. So you can carry out an Ohm's Law calculation for each resistor and add them up to get your total circuit voltage and resistance – occasionally useful if the bulbs can be unscrewed.

In **parallel circuits** adding resistors reduces the total resistance; but the voltage remains constant.

Kirchhoff's current law tells us that the current entering a junction equals the current leaving, so:

Total Current = I1 + I2.............

As I = V/R,

V/Total R = V/R1 + V/R2...........

If we divided throughout by V

1/Total R = 1/R1 + 1/R2............

Parallel circuits are used for lighting circuits as all lamps will give equal brightness and if one fails the others will remain lit. The lamp that is out is obvious and so no need for calculations, however it can be useful with a voltmeter to trace faults on items like printed circuit board dashboards or rear lamp assemblies.

ELECTRICAL AND ELECTRONIC COMPONENTS

Battery – on a race car these are normally gel batteries, that is the electrolyte is a gel (think of jelly) so that when inverted the electrolyte will not leak out. The plates are calcium based so that no topping up of the electrolyte is needed.

Once the electrolyte is added to the battery it will start to have a chemical effect on the plates, in other words the plates will start to deteriorate. Under these conditions batteries have a limited life, so they are usually stored without electrolyte. To make up a gel battery the electrolyte is added to the cells in a liquid form, then put on a slow charge for six or seven hours so that the liquid turns into gel. The charging must be done in one session with the battery off the vehicle using a variable voltage charger.

Gel batteries only have a limited ampere/hour capacity and discharge rate. That is they are only suitable for powering electrical systems on the running vehicles, not starting. There are some exceptions, such as starting motorcycle engines. So to start a race car it is normal to have a battery on a trolley which plugs into a socket on the vehicle, or a plug-in starter pack. The jack plug may be positioned left, right or rear depending on circuit direction. Most circuits are clockwise, so a left-hand drive car is easier to drive, and a socket on the right the keeps the mechanic out of the traffic.

FIGURE 7.3
Dashboard of a racing car

The best way to test a gel battery is by measuring the voltage, do not use a high-rate discharge tester or other heavy load devices.

Rally cars and more robust race cars may use heavy duty lead-acid batteries. In the case of rally cars the battery must be mounted in a separate sealed battery box in case of a roll; this will prevent electrolyte from spilling on the driver and the navigator. On rear wheel drive cars it is a good idea to mount the battery near the rear axle to improve traction and weight distribution.

Ampere hour capacity (AmpHr) is the number of amps a battery will supply for one hour. However, the normal test is over a period of 20 hours. So a 40 AmpHr battery will supply 4 Amps for 20 hours. Rally cars use at least 1.5 times the recommended capacity – the extra weight does not outweigh the risk of a flat battery following stalling on a night stage.

Reserve capacity is the time in minutes that a battery will supply 25 Amps at 25 degrees C. Typically this figure is about 60 minutes. This figure may replace the AmpHr capacity.

Cold cranking is a measure of the ability to provide a current of 170 Amp at –18 degrees C (0 degrees F) at greater than 8.4 Volt. There are SAE, BSI and DIN standards for this. Generally if you take the race vehicle out for February testing and it will not turn the engine over, it is an internal battery problem. If before you re-fitted the battery you had slow charged it overnight and cleaned and put petroleum jelly on the terminals, the plates have sulphated – that is the filler paste has turned solid and will not pass electricity. Do not waste testing time (and track costs) messing with the battery, fit a new one.

Sensors are used to gather information for transmission to the various ECUs (Table 7.2).

TABLE 7.2 Sensors and Locations

No.	Sensor	Location	Comment
1	Acceleration/ deceleration	Scuttle or chassis	For data logging
2	Ambient temperature	Front valance or mirror	May 'ping' warning on dashboard
3	Cabin temperature	Up to four foot wells	With automatic temperature control
4	Coolant temperature	Radiator and/or block	To sense cold starting, or re-starting with electronic engine management
5	Deceleration	Front valance and/or scuttle	Two sensors with SRS
6	Door open	'A' post	Also for boot and bonnet
7	Engine knock	Engine block	Piezo electric
8	Engine speed		Speed counter and TDC locator may be separate
9	Fuel pressure	Fuel system	High pressure side
10	Gear selector	Gear lever	Or gear box
11	Induction vacuum	Inlet manifold	May have pipe to ECU
12	Lambda	Exhaust front pipe	May have before and after catalytic converter – two for each pipe
13	Lap time	Instrument panel	Signal from start/ finish line
14	Lateral acceleration	Scuttle or chassis	For data logging
15	Oil pressure	Main oil gallery	May also be on cooler
16	Oil temperature	Sump or oil cooler	May have thermostat on cooler circuit
17	Outside light level	Windscreen	
18	Rain	Windscreen	
19	Road speed	Gearbox output shaft	
20	Seat belt	Connector	
21	Throttle position	Throttle butterfly	
22	Transponder	Upper right of roll cage	For race control

(Continued)

TABLE 7.2 *(Continued)*

No.	Sensor	Location	Comment
23	Turbo pressure	Turbo outlet	
24	Tyre pressure	Valve and wheel arch	
25	Vehicle direction	As road speed	
26	Vehicle lights	Lamps	
27	Wheel speed	Suspension upright brake disc	For ABS and tyre pressure

All sensor faults are found using the system diagnostic socket. As they are mainly solid state digital electronics they are robust and unlikely to fail unless abused or misused. The technician is advised only to test them following the manufacture's methods and procedures. Do not apply any 12 V supply or use a multimeter on ohm setting, as this is likely to damage the sensor. Operating voltage is in the region of 2V to 5V; with the current in low milliamps.

ECU – the electronic control unit is based one or more integrated circuit (IC) devices (referred to as chips). The ICs are about 40 mm by 20 mm (3/4 inch by 1/2 inch) and contain detailed circuits. There are a number of ICs in the ECU with appropriate circuits and a connector. The ECU typically operates at 5V.

FIGURE 7.4
Lap information indicator in the ideal position for the driver

There are a number of different types of ICs. Random Access Memory (RAM), Read Only Memory (ROM), Electronically Programmed Read Only memory (EPROM) and Electronically Erasable Read Only Memory (EEPROM).

The principle of operation is that of logic. The IC gathers information from a number of sensors and makes an on or off decision. The basis of digital electronics is 0 or 1, off or on. The simplest system uses logic gates. For ignition and fuelling it is normal to use look-up tables. In this case the EPROM is programmed to give a specific output for given inputs rather than making a series of on–off decisions. There may be ECUs for:

- fuel only
- ignition only
- ignition and fuel (engine)
- transmission
- braking only
- chassis – braking and other system controls
- data logging.

THINK SAFE

Only test ECUs using the diagnostic socket.

Actuators (see Table 7.3) are components which carry out a mechanical function in response to an electrical signal. As cars increasingly move to drive by wire, the use of actuators increases.

Actuator faults are tested through the system diagnostic socket, as they work at low voltages with low currents. Do not test by applying battery voltage unless the manufacturer's procedure advises you to do this.

RACERS NOTE

When on-event you may not have access to a full range of fault code reader, experience should suggest to you which sensors and actuators are likely to fail, so it is a good idea to carry spares which can be substituted quickly.

CABLES AND WIRING

The purpose of the electrical cables is to get electrical energy from one part of the vehicle to another. On a competition vehicle this needs to be done with the minimum amount of weight and the least loss of power.

TABLE 7.3 Actuators

No.	Actuator	Location	Type	Comment
1	ABS	Scuttle or chassis	Spool valves	From ECU
2	Choke control	Side of carburetter	Stepper motor	
3	Cruise control	Throttle control and gear change		Constant speed with maximum fuel economy
4	Fuel cut-off	Fuel line	Inertia switch	
5	Gear change	Gearbox	Solenoid controlled valves	
6	Solenoid	Scuttle or chassis	Single movement of armature	Inertia type starter motor
7	Solenoid	Starter motor body	Two stage movement	Pre-engaged starter motor
8	SRS	Air bag and anti-submarine device	Pyrotechnic	
9	Stepper motor	Carburetter, light position	Motor moves in 15 degree steps (or as programmed)	Precise control of shaft movement
10	Temperature control	Air conditioning controls	Solenoid or stepper motor control of valves	
11	Throttle movement	Throttle butterfly	Stepper motor	

RACER NOTE

The most common cause of electrical system failure is a dry joint, where the cable does not join the connector properly and fails to pass a current.

A number of different types of cables and connectors are used. The simplest and cheapest connectors are snapping connect plastic ones – used on production vehicles with either copper or aluminium cable. The cables may be soldered to the connectors, or crimped on the end. Soldering gives lower electrical resistance and can be used with either copper or aluminium cable. Crimping is cheaper and quicker; but likely to separate if subjected to vibrations, and if wet, corrosion between the cable and the connector may prevent the flow of current.

F1 cars use small screw-together connectors to give weatherproof sealing and low risk of separating. The cable is very low resistance so that a complete wiring loom is about the same diameter as one large cable on a standard saloon. The cable and connectors may be gold coated. Gold is a noble metal and will not corrode under most conditions as well as being the best electrical conductor. F1 and similar professional wiring harnesses with connectors may cost the same as a complete Formula Ford car.

RACER NOTE

Noble metals include: Gold, Silver, Platinum, Iridium, and Rhodium. You can wear these as jewellery without your skin going green.

To make up a **loom** for a racing car the best plan is to set out the components on a clean surface – or dummy components, and measure the relevant distances, remember changes in height too. You can then make up a CAD drawing from your measurements. This drawing can be printed out full size as a template. Or, using sheets of plywood make a dummy of the chassis and place pins in it where the cables are going – remember the changes in height too. Then using cable off reels, thread it out and bind it into looms. For this you can use a wide tape or a shrink wrap material. Make the loom as small and tight as possible. Leave free ends sufficient for termination.

The choice of cables will be dictated by the current that they need to carry – generally apply a safety factor of two. Colours should follow a typical colour code. For example: supply brown, post-ignition white with tracer colour, earth black, auxiliary green with tracer colour, lighting red and or blue. Data logging – follow the diagram supplied with the data logging kit.

Multiplexing is a system of transmitting data in a shared environment, at the same time connecting up the components in a circuit. In computer terminology it is called a Local Area Network (LAN). There are several different types of multiplexing systems, as you would expect from different manufacturers. Examples are:

- controller area network – CAN-DATABUS (Bosch)
- body electronics area network – BEAN (Toyota/Denso)
- universal asynchronous receiver transmitter – UART
- audio visual communications – local area network – AVC-LAN.

Most systems work with a twisted pair of cables (like telephone systems) to reduce the possibility of interference. The working voltage is generally 5V with a band rate of 10 to 100 Kbits/second. All the ECUs are connected to form a daisy chain. Sensors and actuators are connected to the network.

Fault finding is by fault code reader in the first instance. Damaged cabling will not show a fault code and will be found by inspection, usually after a serious incident or an off. Looms cannot be repaired and must be replaced in complete sections – they are very expensive.

ELECTRONIC BRAKE CONTROL

Anti-lock braking systems (ABS) and anti-skid control systems (electronic stability control) work through the same ECU. The ECU constantly monitors the speed of each wheel through the wheel speed sensors. If under braking one wheel slows more than the other, then it is likely that the greater slowing wheel is on ice, or other slippery surface. If it were to lock then the vehicle would skid. Within about 2 milliseconds (0.002 s), faster than you or I can blink, the ECU sends a signal to the ABS actuator telling it to reduce the brake fluid pressure to that wheel; this prevents the wheel from skidding. This activity is continuous all the time that the brake pedal is depressed. You can feel the pedal going up and down as this is happening, the dashboard warning light may illuminate. ABS does not work on soft snow, or gravel, as there is no positive traction with these surfaces. In these instances you should switch the ABS off. **Electronic Stability Control** systems prevent wheel spin by doing exactly the same as the ABS; but in the opposite way. That is if one wheel spins more than the other the brake is applied to the spinning wheel to slow it down. This also works when cornering to prevent the vehicle from rolling over.

IGNITION SYSTEM

Since about 1990 all production cars have been fitted with electronic ignition systems. On classic cars before this date you may find mechanical contact breaker points in a distributor with coil ignition. In the 1960s DIY electronic ignition became available and was used extensively on motorsport vehicles.

RACER NOTE

The conventional cb points and coil ignition was invented by Dr Kettering working on the racing Cadillac of 1905 – becoming almost universal for nearly 90 years.

Spark plugs – race plugs differ from standard road plugs in that they will withstand much higher temperatures and pressures for longer periods. The spark plug in a race engine must be capable of providing a spark to ignite the mixture at almost double the pressure of a road vehicle engine, at twice as many times per minute and hence about 400 degrees C hotter.

Before replacing any spark plugs check:

- reach
- diameter
- thread type
- seating or sealing type
- socket fit
- electrode type
- heat range.

RACER NOTE

Engine speed: standard 8,000 rpm; race over 18,000 rpm.
Compression ratio: standard 9 : 1; race 16 : 1.

Ignition fault finding is mainly down to the fault code reader. However, if you are **tuning** a competition engine you may wish to investigate a few of the areas and make adjustments. Some of the adjustments may mean altering the ignition IC programme – see **chipping**.

- Plug temperature – plugs will run hot if the mixture is too weak, you can spot this if you look at the insulator core. If it is too hot it will look a very light brown, or white, and look burnt. Dark, or black, means too rich or oily. Look out for bits stuck to it too. Retarded timing also causes overheating.
- Burn time – you can get this from your engine test bench oscilloscope or digital test kit.
- Dwell angle.
- HT value – most important is that they should all be equal. If one is too high or too low, suspect a plug fault too.
- Ignition timing – especially at high revs.

FUELLING SYSTEM

From the point of electronics, fuel systems have had a number of different systems up to the current electronic fuel injection (EFI). As a motorsport technician you will come into contact with many old systems as well as the current ones. Some of the more common ones involving electronics:

Electronic carburetter – developed by SU, uses a stepper motor to move the jet to control the mixture for cold starting. The mechanism between the stepper motor and the jet is subject to wear, so many were converted to a mechanical choke.

Single-point fuel injection – this is a step up from the carburetter. It is fully electronically controlled through the engine ECU. It has two fuel outlets, one for normal running and one which cuts in on cold start. Maintenance is mainly changing the air filter and the petrol filter. However, the throttle potentiometers are prone to failure – this causes intermittent surging. The fault can easily be verified by the fault code reader.

Electronic fuel injection (EFI) – by this we mean multi-point, either individually supplied by fuel, or **common rail** which is almost universal. The pressure pump sends fuel to the injectors at a pressure of between 4 and 8 bar (60 to 110 psi) depending on the vehicle and the regulator setting. The amount of fuel going into the cylinder depends on this pressure (which is pre-set) and the amount of time that the injectors are open. The longer the injectors are open the greater the amount of fuel that will be injected. The ECU controls the open time depending on engine requirements, these include:

- load conditions
- engine speed
- engine temperature
- ambient temperature
- lambda feed back
- throttle position.

This information is fed to the ECU by the sensors, see **ECU** and **Sensors**.

The injectors are held open by a **duty cycle**, not a continuous current. The time period is between about 4 and 8 milliseconds (0.004 s to 0.008 s). Doubling the time at a constant pressure will double the amount of fuel injected.

The flow of fuel is also controlled by the size of the injector orifice. Standard injectors are often referred to as 'grey' because of their colour code. For high performance engines 'green' injectors are used – these have a bigger orifice and can therefore deliver more fuel.

Fault finding on EFI systems should start with the fault code reader. The air filter and the fuel filter should be changed regularly. The injectors are tested by removing them and observing their spray pattern on a test rig. The main problems are usually related to sensor failure – typically the throttle position sensor because it moves and wears, or the lambda sensor because it gets hot and deteriorates.

RACER NOTE

Ultrasonic cleaning using a system such as ASNU will save the cost of new injectors.

CHIPPING

The operation of the ignition and the fuelling systems depends on the **mapping** of the ICs (chips). For race and rally cars different ICs are required to suit the application. Fitting a specially set up IC to the ECU is called **chipping**. It is possible to buy ready set up ICs for most popular applications (Ford, BMW, Peugeot, Vauxhall and similar); and to set up the chip for an individual application. When a chip is changed it is normal to check the power before and after on a rolling road dynamometer (dyno). If an individual chip is being set up then expect to spend two or more days on the dyno with a lot of test equipment.

The chip is set as a contour map. Parameters for engine speed on the x axis and load on the y axis are converted into ignition setting and fuelling settings on the z axis. This is read on a laptop computer as a look-up table. Each square on the table has a value. Reading the table rows from left to right gives the figures for increasing load, going down the columns the engine speed is increasing. The figures in the fuelling table correspond to the opening times of the injectors – figures such as 0.004 s (4 milliseconds). With a suitable interface between the laptop and the chip, and access software, you can alter these figures and check the changes to the power output on the dyno. As you can see, if the table has 40 rows and 20 columns that is 800 individual boxes to check.

LIGHTING

Headlamps and rear lamps used in Europe must comply with EU regulations and carry the 'e' or 'E' mark. In the UK they must dip left; in continental Europe they must dip right.

RACER NOTE

To use some UK cars at night in continental Europe it is usually necessary to place specially shaped black tape over the lamps to prevent dazzle to the oncoming cars.

All complying lamps, head and tail indicators, stop and fog carry a set of digits as well as the EU mark. These digits are a code which indicates the position and type of lamp.

FIGURE 7.5
Paulio Racing's LED rear light

FIGURE 7.6
KTM rear quarter

FIGURE 7.7
Audi RS8 rear quarter

FIGURE 7.8
Ducatti 1098 rear view. Can you think of problems the designer had with this set up?

LED lamps have now become popular. Their slow introduction was because of Construction and Use Regulations which required a light filament, but this regulation was withdrawn. LEDs have the advantages of:

- using less power
- being quicker in reaction
- being more robust and longer lasting
- allowing the use of different shapes.

Gas discharge lamps (GDL) – this is a form of arc light. The lamp is about the same size as a normal halogen one made of quartz glass. The high-voltage arc is provided by an electronic control unit from the 12V system. It has become popular with scooter riders as it uses only a low current. The amount of light given is about three times that of a tungsten bulb for the same current usage.

Ultra Violet or **blue lights** are sometimes used as one pair on multi-lamped vehicles. UV light will light up road markings and be almost invisible to oncoming traffic. They do not give back reflection in fog and snow, therefore appearing to see through them. As UV rays can be dangerous a two-layer filter is fitted to remove UVB and UVC – the dangerous ones, leaving UVA. The usual light source is a GDL.

CHAPTER 1 ENGINE

1. To fit wet liners to an open block you should:
   - a. Use your hands
   - b. Use a hammer
   - c. Use a 10 tonne press
   - d. Use your feet
2. Engine oil in a rally car is likely to run at:
   - a. 80 degrees C
   - b. 800 degrees C
   - c. 180 degrees C
   - d. 380 degrees C
3. Cooling system hoses on a race car are made strong to resist:
   - a. Mechanics tools
   - b. Very low temperatures
   - c. Torsional stress
   - d. Hoop stress
4. The ratio which relates the amount of power out of an engine to the amount of petrol used is called:
   - a. Volumetric ratio
   - b. Thermal efficiency
   - c. Compression ratio
   - c. Lambda
5. Antiknock Index is:
   - a. ROM + MON/2
   - b. CR + CV
   - c. HP/Torque
   - d. V1 + R1
6. Joule is the term for:
   - a. Power
   - b. Torque
   - c. Energy
   - d. Temperature
7. If the A/R ratio on one turbocharger is smaller than that of another, the smaller numbered one will:
   - a. Respond more slowly
   - b. Respond at the same rate
   - c. Respond more quickly
   - d. None of the above applies

8. Air density at sea level at NTP is:
   a. 1200 kg/m^3
   b. 1.2 kg/m^3
   c. 120 kg/m^3
   d. 0.12 kg/m^3
9. To measure the air flow through the cylinder head you would use:
   a. A micrometer
   b. A test bench
   c. A flow bench
   d. A vernier
10. Over square and square are terms used to describe:
   a. Valve layouts
   b. Bore–stroke ratios
   c. Con-rods
   d. Crankshafts

Further work

a. When you carry out any work on an engine find out its full specification and make notes about it.
b. Use the calculations in this chapter where possible.
c. Run different types of engines and listen to their exhaust and general running noises – make notes about what you hear.

CHAPTER 2 CHASSIS

1. The difference between the overturning moment and the righting moment is called:
   a. Track
   b. Roll
   c. Castor
   d. Camber
2. The steering angle which gives self-centring after cornering is called:
   a. Castor
   b. Camber
   c. KPI
   d. Ackermann
3. When carrying out a check for toe-out on turns and the outer wheel is turned through 20 degrees, the inner wheel should turn through:
   a. 20 degrees
   b. 22 degrees
   c. 18 degrees
   d. none of these
4. The wheel, hub and suspension upright all form part of the:
   a. Springing medium
   b. Drive train

c. Sprung mass
d. Unsprung mass

5. The difference between the direction of travel of the vehicle and the wheel angle is called:
   a. The camber angle
   b. The fixed earth system
   c. The slip angel
   d. The Ackermann angle
6. The difference between the centre line of the wheel and the centre line of the suspension upright is called:
   a. Toe-in
   b. Toe-out
   c. Scrub radius
   d. Wheel trail
7. The organisation in the UK which issues competition licenses for most branches of motorsport is:
   a. The MSA
   b. The MIA
   c. The IMI
   d. The IMechE
8. Most electronic circuits which are controlled by an ECU can be fault diagnosed with:
   a. A fault code reader
   b. An ohm meter
   c. A volt metre
   d. An ammeter
9. The position of the centre of gravity can be found using:
   a. A rule
   b. A hoist
   c. A spring balance
   d. A set of corner weights
10. Lap time, lateral acceleration and throttle position are all likely to be found on:
    a. A laptop
    b. A data acquisition system
    b. An engine ECU
    c. A chassis ECU

Further work

a. Carry out any of the tasks mentioned in this chapter.
b. Look at a single seat race car and locate all the components which we have discussed.
c. Drive as many different cars as you can and compare their handling with one of you class mates.

CHAPTER 3 TRANSMISSION

1. The date of manufacture of a transmission in Julian form for 31st January 2009 is written as:
   a. 3119
   b. 3109
   c. 9031
   d. 9311
2. The maximum torque which can be transmitted by a torque is defined by the formula:
   a. S P μ R
   b. Ma
   c. 2pr
   d. SPD
3. In an automatic gearbox the rear pump is used mainly when:
   a. The car is cruising in top gear
   b. The car is braking heavily
   c. The car is accelerating hard
   d. The car is travelling slow in traffic
4. The most fuel efficient way to drive an automatic car on the motorway is to select:
   a. Sport mode
   b. Lock-up – L
   c. Cruise control
   d. Drive 2
5. The first check to make on a suspected faulty automatic transmission is:
   a. Plug in the code reader
   b. Carry out a stall test
   c. Carry out a full pressure test
   d. Check the fluid level and casing for leaks
6. In the formula $R_a = C_d A V^2$ the term C_d refers to:
   a. The drag coefficient
   b. The sound system type
   c. The car door
   d. The air mass
7. The Tractive Effort (TE) must be greater than:
   a. The vehicle mass
   b. The road wheel friction
   c. $R_a + R_r + R_g$
   d. S P μ R
8. Many race and rally gearboxes have gear teeth which are:
   a. Straight cut
   b. Double helical
   c. Hypoid bevel
   d. Straight bevel

9. The simplest way to change the overall gear ratio is to:
   a. Change the gearbox
   b. Change the final drive
   c. Change the clutch
   d. None of the above
10. Gradient resistance equals:
   a. S P μ R
   b. $R_a + R_r + R_g$
   c. sinG W
   d. Ma

Further work

a. Investigate the gear change mechanism on a single-seat race car – make a drawing of the joints.
b. Calculate the overall gear ratio for a car of your choice.
c. From maximum engine speed, overall gear ratio and tyre rolling radius calculate a vehicle's maximum speed.

CHAPTER 4 INSPECTION

This chapter requires you to complete the test sheets; guides to the answers are given. The further extension work is to progress your career to the next level of inspection, which is you should investigate the steps your need to take and the competencies that you need to gain to become:

- an MSA or other motorsport organising body scrutineer
- an MOT Tester.

CHAPTER 5 OVERHAUL

1. Original parts are often referred to as:
   a. OEM parts
   b. Pattern parts
   c. Re-manufactured parts
   d. Spare parts
2. When stripping a part that you have not seen before it is a good idea to:
   a. Throw it in the parts tray
   b. Throw it in the rubbish bin
   c. Photograph it before you remove it
   d. Photograph it before you re-fit it
3. To make it easier to overhaul an engine it is good to use:
   a. New tools
   b. An engine stand
   c. An axle stand
   d. Old tools

4. A motorsport cluster is:
    a. Where a group of motorsport companies work close together
    b. A motor club
    c. Some sort of tool
    d. A convoy of transporters going to an event
5. The first task before starting to work on a damaged rally car is likely to be:
    a. Removing the damaged part
    b. Pressure washing the area
    c. Rolling it on its side
    d. Ordering the new parts
6. When re-building an engine, the action you should take with high tensile cylinder head bolts is:
    a. Re-tap the treads on them
    b. Clean them with wire wool
    c. Replace them all
    d. None of the above
7. To get the most power out of an engine which you are re-building you must ensure that all the parts:
    a. Are to exact tolerance
    b. Fit together tightly
    c. Fit together some free play
    d. Are all lubricated
8. You would measure a crankshaft for wear with:
    a. An internal micrometer
    b. A vernier gauge
    c. Callipers and a rule
    d. An external micrometer
9. If a crank pin measures different diameters when taken at 90 degrees to each other, it is said to have:
    a. Ovality
    b. Taper
    c. No wear
    d. The wrong bearings
10. To ensure that a proper record of any work done on a race car is recorded you must:
    a. Tell the owner
    b. Tell the driver
    c. Enter it in the log
    d. Record it on the data logger

Further work

a. Check the overall procedure for your favourite vehicle engine.
b. Investigate your company's procedure for blue printing.
c. Find the tolerances for pistons and bores of an engine which you like.

CHAPTER 6 MODIFICATIONS

Your work for this chapter is to gain some product knowledge about the type of vehicle which either you work on, or particularly interests you. We hope that both are the same; but not everybody is so lucky.

Many vehicle manufactures produce modification, or accessory kits. Others have a number of different model variations. Specific product knowledge is very important; you can use it to build the fastest vehicle, or one for a special purpose. You can use this information to advise customers and hopefully make a greater profit.

CHAPTER 7 ELECTRICAL AND ELECTRONIC PRINCIPLES

1. $P = I^2R$ is a formula used to calculate:
   a. Power
   b. Torque
   c. Resistance
   d. Current flow
2. To calculate total resistance use:
   a. R1 + R2..............
   b. V1 + V2..............
   c. A1 + A2...............
   d. A1 + V1 + R1.......
3. The non-dazzling light that is good in fog and rain with a coloured hue is called a:
   a. Red light
   b. Blue light
   c. Yellow light
   d. Green light
4. The technical name for a 'chip' is:
   a. Actuator
   b. Integrated Circuit (IC)
   c. ECU
   d. Thermistor
5. The current leaving a junction equals the current entering it is a statement of who's law?
   a. Pascal
   b. Newton
   c. Kirchhoff
   d. Bugatti
6. The cold cranking test on a battery is carried out at:
   a. 40A at STP
   b. 170A at – 18 deg C
   c. 40A at NTP
   d. 170A at 80 deg C

7. Knock sensors work on the vibrations from incorrect combustion causing a tiny current to send a signal to the ECU, they operate on a principle known as:
   a. Piezo electric
   b. Self-induction
   c. Self-exciting
   d. Capacitance
8. The best conductor of electricity is gold which is also:
   a. Cheap to buy
   b. A noble metal
   c. A base metal
   d. Not used on motorsport vehicles
9. AVC-LAN and CAN-DATABUS are both types of:
   a. Multiplex systems
   b. ICs
   c. EPROMS
   d. Lighting devices
10. The ABS system responds to signals from the sensors in:
   a. 0.002 s
   b. 0.020 s
   c. 0.200 s
   d. 2.000 s

Further work

a. Sketch out the inputs and outputs to the ECU on any car in the workshop.
b. Investigate the spark plugs used in one of the vehicles which you work on, find out the following: reach, length, heat range, gap, socket size, operating voltage.
c. Test drive several cars with different lighting systems in the dark – write up some notes about your experiences.

ANSWERS

Question No.	Chapter 1	Chapter 2	Chapter 3	Chapter 5	Chapter 7
1	a	b	c	a	a
2	c	a	a	c	a
3	d	b	a	b	b
4	b	d	c	a	b
5	a	c	d	b	c
6	c	c	a	c	b
7	c	a	c	a	a
8	b	a	a	d	b
9	c	d	b	a	a
10	b	b	c	c	a

BIBLIOGRAPHY

Bonnick, Allan. *Automotive Computer Controlled Systems,* Butterworth Heinemann.

Chowanietz, Eric. *Automobile Electronics.* Newnes – Butterworth Heinemann.

Gillespie, Thomas D. *Fundamentals of Vehicle Dynamics,* SAE International.

Glimmerveen, John H. *Hands-on Race Car Engineer*, SAE International.

Hartley, John. *The Fundamentals of Motor Vehicle Electrical Systems,* Longman.

Heisler, Heinze. *Advanced Engine Technology,* Edward Arnold.

Heisler, Heinze. *Advanced Vehicle Technology,* Edward Arnold.

Livesey, WA. *Motor Vehicle Studies,* Cassells.

Livesey, WA. *Vehicle Mechanical and Electronic Systems,* Institute of the Motor Industry.

Livesey, WA and Robinson, A. *The Repair of Vehicle Bodies,* Butterworth Heinemann.

Milliken, William F and Milliken, Douglas L. *Race Car Vehicle Dynamics*, SAE International.

Newton, K, Steeds, W and Garrett, TK. *The Motor Vehicle,* Butterworth Heinemann.

Many of these books are available from the Institute of the Motor Industry's on-line book store www.motor.org.uk/store

The student is also directed to serious professional motorsport magazines such as: *Autosport, Track and Race Cars,* and *Motorsport.*

GLOSSARY

This section defines a number of the words and phrases used in motorsport engineering, including some of the specialist racer and enthusiast vocabulary and jargon

'An off' when you come off the circuit unintentionally into an area where you are not supposed to – if you are lucky it is just grass
'Esses' one bend followed by another
'O' rings rubber sealing rings
$^1/_4$ mile quarter-mile drag racing strip
$^1/_8$ mile eighth-mile drag racing strip
24 hour a race lasting 24 hours, the winner is the one covering the greatest distance, usually there are different capacity based classes. For low budget racers there is an event at Snetterton for Citroen CVs and one in Sussex, near a pub, for lawn mowers
Acceleration rate of increase of velocity
Accessories anything added which is not on standard vehicle
Ackermann steering set up to prevent tyre scrub on corners
Add-ons something added after vehicle is made
Adhesion how the vehicle holds the road
Air-bags SRS – bags which inflate in an accident
Alignment position of one item against another
Alloy mixture of two or more materials; may refer to aluminium alloy, of an alloy of steel and another metal such as chromium
Ally aluminium alloy
Anti-roll bar suspension component to make car stiffer on corners
Atom single particle of an element
Back-fire when the engine fires before the inlet valves are closed – sending flames and gas out of the inlet
Barrel cylinder barrel – usually refers to motorcycle engine
BDC bottom dead centre
Bench working surface; also flow bench and test bench
Beta version test version of software, or product
Birdcage tubular chassis frame which resembles a bird cage
Block and tackle used to lift engines
Block cylinder block – a number of cylinders in one piece
Bonnet engine cover (front engined car)
Bore internal diameter of cylinder barrel
Brooklands first purpose built racing circuit at Weybridge in Surrey with banking and bridge
Cabriolet four seater convertible body
Carbon fibre like glass fibre but uses very strong carbon based material
Cetane resistance to knock of diesel fuel
Chicane sharp pair of bends – often in the middle of a straight
Chocks tapered block put each side of wheel to stop the car rolling
Circuit race circuit
Clerk of Course most senior officer at a motorsport event – person whose decision is final, although there may be a later appeal to the MSA or FIA
Code reader reads fault codes in the ECU of the particular system

Composite material made in two or more layers – usually refers to carbon fibre, may include a honeycomb layer
Compression ratio ratio of combustion chamber size to cylinder bore
Con rod connecting rod
Condensation changes from gas to liquid
Contraction decreases in size
Corrosion there are many different types of corrosion, oxidation/rusting are the most obvious
Cubes cubic inches – USA term for size of engine, the saying is, 'there is no substitute for cubes'.
Cushion section of seat to sit on
Dashboard instrument panel
Density relative density also called specific gravity
Diagnostic machine equipment connected to the vehicle to find faults
Dive vehicle goes down at front under heavy braking
DOHC double overhead cam
Double 12 a 24 hour event divided into two 12 hours – day time only with parc ferme in the evening
Double 6 another name for a V12 engine
Drag racing two cars racing by accelerating from rest on a narrow drag strip
Engine cover cover over engine (usually refers to rear engined car)
Epoxy resin material used with glass fibre materials
Evaporation changes from liquid to gas
Event organiser person who organises the race or other event
Expansion increases in size
Fast back long sloping rear panels
Fender USA for mudguard or wing
FIA international motorsport governing body
Flag marshal marshal with a flag
Flag chequered flag, black flag, and red flag
Flow bench used to measure the rate of flow of inlet and exhaust gas through a cylinder head
Foam material used to make seats and other items
Force mass times acceleration
Formula car car built for a specific formula, usually refers to open-wheeled single-seaters such as Formula Ford or Formula Renault
Friction resistance of one material to slide over another
Frontal area (projected) area of front of vehicle
Gelcoat a resin applied when glass fibre parts are being made – it gives the smooth shiny finish
Glass fibre light weight mixture of glass material and resin to make vehicle body
GT Grand Turisimo (Italian for Grand Touring) first used on a Ferrari with two passenger doors and a rear luggage opening
Hatchback four seater with rear upward opening rear
Heat a form of energy
Hill climb individually timed event climbing a hill
Hoist used to lift vehicles, may be two-post or four-post
Hood USA for bonnet
Inboard something mounted on the inside of the drive shafts such as inboard brakes, usually lowers unsprung weight
Inertia resistance to change of state of motion – see Newton's Laws, inertia of motion and inertia of rest
Intercooler air cooler between turbocharger and inlet manifold – to cool incoming air for maximum density
Kevlar fibre super strong material, often used as a composite with carbon
Le Mans Series 24 hour sports car races across the world – many in USA, organised by ACO

Le Mans 24 hour race for sports cars, prizes for furthest distance covered and best fuel consumption organised by Automobile Club de l'Ouest (ACO) (West France Auto Club)
Marshal person who helps to control an event
Mass molecular size, for most purpose the same as weight
Metal fatigue metal is worn out
Molecule smallest particle of a material
Monza World's second banked race track in Italy, copy of Brooklands
Motocross off-road circuit event (motorcycles)
MSA Motor Sport Association
Newton's Laws First Law – A body continues to maintain its state of rest or of uniform motion unless acted upon by an external unbalanced force
Second Law – the force on an object is equal to the mass of the object multiplied by its acceleration (F = Ma)
Third law – to every action there is an equal and opposite reaction
Nose cone detachable front body section covering front of chassis – may include a foam filler for impact protection
Octane rating resistance of knock of petrol
Off-roader vehicle for going off-road; or off-road event
OHV overhead valve
Open wheeler race (circuit) car with no wheel covering
Original finish original paint work, usually with reference to historic or vintage cars
Outboard something mounted on the outside of the drive shafts such as brakes, usually increases unsprung weight
Oxidation material attacked oxygen from the atmosphere; aluminium turns into a white powdery finish
Paddock where teams and vehicles are based when not racing
Parc ferme area where competition vehicles are left and cannot be worked on or prepared, usually between race rounds or rally stages
Parent metal main metal in an item
Pit garage garage in pit lane; also just garage
Pit lane lane off the circuit leading to the pits
Pit wall protecting the pit lane from the circuit
Pit place for preparing/repairing/re-fuelling the vehicle at the side of the circuit
Pot another name for cylinder
Power work done per unit time, HP, BHP, CV, PS, kW
Prepping preparing the vehicle for an event
Prototype first one made before full production
Rally-cross off-road circuit event (rally cars)
Regs racing regulations
Ride height height of vehicle off the road, usually measured from hub centre to edge of wheel arch
Rings piston rings
Roll bar frame inside vehicle which is resistant to bending when vehicle rolls over – safety protection for occupants
Rust oxidation of iron or steel – goes to reddish colour
Saloon standard four seater car body
Scrutineer person who checks that a vehicle complies with the racing regulations, usually when scrutineered the vehicle has a tag or sticker attached
Side valve when the valves are at the side of the engine (old engines and lawn mowers)
Single seater see formula car
SIPS side impact protection system – door bars (extra bars inside doors)
Skid vehicle goes sideways – without road wheels turning
Speed event any event where cars run individually against the clock
Spine backbone-like structure
Sprinting individually timed event starting from rest over a fixed distance
Sprung weight weight below suspension spring

Squab back of seat – upright part
Squat vehicle goes down at back under heavy acceleration
Squeal high pitch noise
Squish movement of air fuel mixture to give better combustion
SRS supplementary restraint system – air bags
Stage rally when the event is broken into a number of individually timed stages, the vehicles start each stage at pre-set intervals (typically 2 minutes)
Stall involuntary stopping of engine
Steward a senior officer in the organisation of the motorsport event
Straw bales straw bales on side of track for a soft cushion in case of an off
Stress force divided by cross-sectional area
Strip drag racing strip
Stripping pulling apart
Stroke distance piston moves between TDC and BDC
Swage raised section of body panel
Swage line raised design line on body panel
TDC top dead centre
Temperature degree of hotness or coldness of a body
Test bench test equipment mounted on a base unit
Test hill a hill of which the gradient increases as the top approaches, originally the test was who got the furthest up the hill
Tin top closed car with roof
Torque turning moment about a point (Torque = Force × Radius)
Transporter vehicle to transport competition vehicles to events
Tub race car body/chassis unit made from composite material
Tunnel inverted 'U' section on vehicle floor, on front-engined rear-wheel drive cars it houses the propeller shaft – propeller shaft tunnel
Turret USA for vehicle roof
Tyre wall wall on side of track built from tyres – giving a soft cushion in case of an off
Unsprung weight weight below suspension spring
Velocity vector quality of change of position, for most purposes the same as speed

WEBSITES

Please remember that sites can change over night, if you use any websites in your assignments it is good to credit them like those below and add the date viewed at the end of the citation:

Firm or person, subject viewed, website address, date viewed

American Petroleum Institute, oil specification www.apicj-4.org/engineoilguide

AP Racing www.apracing.com

Autosport Show at NEC, Birmingham in January each year www.autosport-international.com (no excuses, get to the show)

Beta Tools, tool manufacturer www.beta-tools.com

British Motor Heritage Ltd http://www.bmh-ltd.com/

Burlen Services, carburetter spares www.burlen.co.uk

Demon Tweeks car and motorcycle competition parts and equipment www.demontweeks.com

Hot rod racing www.nhra.com

Institute of the Motor Industry, professional body www.motor.org.uk

NASCAR racing www.nascar.com

Performance Racing Industry, magazine and annual show in USA www.performanceracing.com (well worth flying to, they organise lots of events during show week, the author won the model racing there one year)

Safety Testing www.euroncap.com

USA car safety www.safercar.gov

USA Government – loads of information www.usa.gov

USA highway safety www.nhtsa.gov

Websites move fast and frequently – for an up to date list of contacts and hyperlinks the reader is directed to the Motorsport Institute, motorsport information site: www.motorsportinstitute.org or www.Livesey.US

VEHICLE MAINTENANCE AND REPAIR LEVEL 3 MOTORSPORT ROUTE

This teaching programme sets out the knowledge and understanding necessary for Candidates to complete the **Level 3 Diploma.** It assumes that Candidates have covered all the knowledge and understanding requirements that are set out for Level 1 & Level 2. Where there is any perceived overlap between Level 2 and Level 3 aspects centres should ensure that the subject matter has been developed in sufficient depth to enable Candidates to conduct diagnosis and fault rectification on complex vehicle systems.

The Level 3 programme is broken down into **four sections** to include the four main vehicle areas designated by the National Occupational Standards developed by Automotive Skills.

- Advance Chassis Technology
- Advance Engine Technology
- Advanced Transmission Technology
- Advanced Electrical & Electronic Technology

Each centre will have its own approach to the delivery of the programme which will be affected by a number of factors including Candidates' mode of delivery (part-time day, block release, full-time), Candidates' background and experience, local circumstances and requirements. Centres may wish to integrate or modify the programme to suit their own particular needs and circumstances.

Centres should also note that the total teaching hours shown in this programme are considerably less than the sum total of the individual unit hours. This is because there is often overlap in the content between individual units. However, when units are grouped together and delivered within a coherent programme these overlaps are removed. If units are delivered on a stand-alone basis the guided learning hours shown on the front of each unit should be noted and the delivery hours adjusted accordingly.

If the candidate has achieved the Level 2 Technical Certificate they do not need to revisit units G1, G2 and G3 at Level 3.

Unit numbers are subject to change, for the purposes of this book the following Unit Numbers are used:

- Engine – MSM07
- Chassis – MSM08
- Transmission – MSM13
- Inspection – MSM06
- Overall – MR11G
- Modifications – MR15

NB: Electrical and Electronic Principles is taken from both Engine and Chassis Units

ADVANCED CHASSIS TECHNOLOGY

The specific and appropriate Health & Safety precautions and legislative requirements should be integrated into the delivery of each aspect of chassis technology to ensure relevance and meaning to Candidates. The legislation and generic provisions and are outlined in units G1, G2 & G3.

Hours	Topic	Technical Certificate Unit Mapping	On-line Assessment	Practical Assessment
6 hours	**Electronic Technology for Chassis Systems:** Operation of electrical and electronic systems and components related to light vehicle chassis systems including ECU, sensors and actuators, electrical inputs, voltages, oscilloscope patterns, digital and fibre optic principles. Interaction between the electrical/electronic system and mechanical components of chassis systems. Electronic and electrical safety procedures.	**MSM08**		
6 hours	**Advanced Braking Systems:** Operation of electronic anti-lock braking, anti-skid control systems. ABS, EBD, layout of systems. Operation of hydraulic and electronic control units (ECU), wheel speed sensors, load sensors, hoses, cables and connectors. Advantages of ABS and EBD braking systems over conventional braking systems. Relationship and interaction of ABS braking systems with and other vehicle systems – traction control. Alternative brake components – 4-pot, increased size components, alternative material (carbon) Synthetic fluids.	**MSM08, MSM15**		
6 hours	**Advanced Steering Systems:** Front/rear wheel geometry: castor, camber, kingpin or swivel pin inclination, negative offset, wheel alignment (tracking), toe-in and toe-out, toe-out on turns and steered wheel geometry; Ackerman principle, slip angles, self-aligning torque, oversteer and understeer, neutral steer. Operation and layout of rear and four-wheel steering. Operation of hydraulic power assisted steering systems: power cylinders, drive belts and pumps, hydraulic valve (rotary, spool and flapper type). Operation of electronic power steering systems (EPS) Quick racks.	**MSM08, MSM15**		

(Continued)

The specific and appropriate Health & Safety precautions and legislative requirements should be integrated into the delivery of each aspect of chassis technology to ensure relevance and meaning to Candidates. The legislation and generic provisions and are outlined in units G1, G2 & G3.

Hours	Topic	Technical Certificate Unit Mapping	On-line Assessment	Practical Assessment
8 hours	**Advanced Suspension Systems:** Pushrod/pullrod linkages. Rose joints. Adjustable axle location. Adjustable ride height. Adjustable shock absorbers (dampers). Self-levelling suspensions. Reasons for fitting ride controlled systems. Operation of a self-levelling suspension system under various conditions. Operation of electronic systems used for control of suspension. Electrical and electronic components. Operation of driver controlled and ride controlled systems. Live hinges and non-metallic components.	**MSM08, MSM15**		
3 hours	**Symptoms and faults associated with braking systems:** ABS, and EBD systems; mechanical, hydraulic, electrical and electronic systems; fluid leaks, warning light operation, poor brake efficiency, wheel locking under braking.	**MSM08, MSM15**		
3 hours	**Symptoms and faults associated with steering systems:** Mechanical, hydraulic, electrical and electronic; steering boxes (rack and pinion, worm and re-circulating ball), steering arms and linkages, steering joints and bushes, idler gears, bearings, steering columns (collapsible and absorbing), power steering system.	**MSM08, MSM15**		
2 hours	**Symptoms and faults associated with suspension systems:** Mechanical, hydraulic, electrical and electronic; conventional, self-levelling and ride controlled suspension systems; ride height (unequal and low), wear, noises under operation, fluid leakage, excessive travel and excessive tyre wear.	**MSM08, MSM15**		

The specific and appropriate Health & Safety precautions and legislative requirements should be integrated into the delivery of each aspect of chassis technology to ensure relevance and meaning to Candidates. The legislation and generic provisions and are outlined in units G1, G2 & G3.

Hours	Topic	Technical Certificate Unit Mapping	On-line Assessment	Practical Assessment
8 hours	**Diagnosis and Testing of Chassis Systems:** Locate and interpret information for diagnostic tests, vehicle, and equipment specifications, use of equipment, testing procedures, test plans, fault codes and legal requirements. Prepare equipment for use in diagnostic testing. Conduct systematic testing and inspection of steering, braking and suspension systems, mechanical, hydraulic, electrical and electronic systems using appropriate tools and equipment including, multimeters, oscilloscope and pressure gauges. Evaluation and interpretation of test results. Comparison of test result and values with vehicle manufacturer's specifications and settings. Dismantling of components and systems using appropriate equipment and procedures. Assessment and evaluation of the operation, settings, values, condition and performance of components and systems. Probable faults, malfunctions and incorrect settings. Rectification or replacement procedures. Evaluation of systems following repair to confirm operation and performance.	MSM08, MSM15		
8 hours	**Setting up, testing and recording motorsport vehicle suspension:** Corner weights, turntables, shock absorber dynamometer, tyre temperature using appropriate manual/software systems, C of G, unsprung weight, ride height.	MSM08, MSM15		
8 hours	**Design, operation, advantages and fitting of add-ons, chassis modifications and safety equipment:** Seam welding, roll cages, glass, wheel arch flares, spoilers, sump guards and under body protection, inside fuel and brake lines, mud flaps, towing eyes, alternative fuel tanks, seat, harnesses, electric cut-off, fire extinguisher systems.	MSM15		
Total Hours 58	**Does not include covering Level 1 & 2 requirements**			

ADVANCED ENGINE TECHNOLOGY

The specific and appropriate Health & Safety precautions and legislative requirements should be integrated into the delivery of each aspect of engine technology to ensure relevance and meaning to Candidates. The legislation and generic provisions and are outlined in units G1, G2 & G3.

Hours	Topic	Technical Certificate Unit Mapping	On-line Assessment	Practical Assessment
6 hours	**Engine Performance:** Meaning of volumetric efficiency, the effect of volumetric efficiency on engine performance, torque and power. Methods used to improve volumetric efficiency, variable valve timing, turbo-charging, supercharging and intercoolers. Construction and operation of turbochargers, superchargers, intercoolers, waste gates and exhaust gas recirculation. Disadvantages of pressure charging induction systems. Layout of multi-valve arrangements, components, operation and drive arrangements. Variable valve timing and multi-valve arrangements and the effect on performance. Construction features and operation of variable valve timing engines and electronic control.	**MSM07, MSM15**		
6 hours	**Engine Combustion:** Flame travel, pre-ignition and detonation. Properties of fuels: octane rating, flash point, fire point and volatility. Composition of carbon fuels (petrol and diesel) % hydrogen and carbon, composition of air: % oxygen, % nitrogen. Combustion process for spark ignition and compression ignition engines. By-products of combustion for different engine conditions and fuel mixtures: CO, CO_2, O, N, H_2O, NOx. Legal requirements for exhaust emissions, MOT requirements, EURO 3, 4 & 5 regulations. Methods used to reduce emissions: low emission fuels, lean burn technology, catalytic converters, closed loop engine management systems.	**MSM07**		

The specific and appropriate Health & Safety precautions and legislative requirements should be integrated into the delivery of each aspect of engine technology to ensure relevance and meaning to Candidates. The legislation and generic provisions and are outlined in units G1, G2 & G3.

Hours	Topic	Technical Certificate Unit Mapping	On-line Assessment	Practical Assessment
6 hours	**Modified cylinder head:** Material, mass, temperature, resistance, ports, finish, valves, machining, gas flowing, gaskets and fasteners. **Valves:** Material, mass, gas flow, wear and temperature resistance, strength, design, gas flow, seat angle, head size, stem diameter. Springs – stiffness, mass, bounce, Guides – material, wear, resistance, friction. Camshaft: profile – high lift, long duration, timing, VVT, materials and surface treatment. Drive train – V belt, toothed belt, gears, chain. Followers – buckets, ratios, rockers, wear resistance, adjustment, roller rockers/followers.	MSM15		
6 hours	**Short block assembly:** Block – material, strength, stiffness, increased capacity, liners, deck height, compression ratio, block ringing, caps, bearing, fasteners, strapping. Pistons – design, balance, clearance, materials, forging, billet, cast, temperature resistance, wear resistance, skirt, groves, mass, rings, design, pre-load, gudgeon pins. Con-rods – mass, design (H, I and solid), balance, piston/rod guided, finish, shot peening, polish, trust effects, bolts/studs, bearings. Crankshaft – materials, surface treatment, design, balance, web, stroked crank, cross drilling, oil ways, bearings, end thrust. Flywheel – material, design (2 piece), ligtening, balance. Lubrication – pump, valves, coolers, dry sump, baffles, oil PU, filter locations. Blueprinting procedures and sealed engines.	MSM15		
6 hours	**Induction and Exhaust Systems:** Inlet and exhaust manifolds – gas flow, length, diameter, layout, siamesing, velocity and pressure calculations. Air filters (cleaners), trumpets, induction tracts and heat shields.	MSM15		
6 hours	**Superchargers and Turbochargers:** Pressure ratios, boost ratios, A/R ratios, waste gates, dump valves, boost control, water cooling, inter cooler, hoses and clips, bearings and service procedures.	MSM15		
3 hours	**Mechanical fuel injection:** Pumps, injectors, throttle systems, pipes, low pressure pumps, pump drive, ram pipes, enrichment devices.	MSM07		

(Continued)

The specific and appropriate Health & Safety precautions and legislative requirements should be integrated into the delivery of each aspect of engine technology to ensure relevance and meaning to Candidates. The legislation and generic provisions and are outlined in units G1, G2 & G3.

Hours	Topic	Technical Certificate Unit Mapping	On-line Assessment	Practical Assessment
6 hours	**Motorsport carburetter systems:** Carburetters, inlet manifolds, throttle linkages, electric and mechanical fuel pumps, fuel pressure regulators, fuel lines and fittings, air filters. **Faults and symptoms in motorsport vehicle carburetter systems:** Poor hot and cold starting, lack of poor performance at low and high speeds, poor idle, hesitation under acceleration, flooding, excessive fuel consumption. **Diagnose faults and carry out adjustments on motorsport vehicle carburetter systems:** Linkages and fittings, mixture strength, idle speed, fuel mixture, pressure regulators, settings, carburetter synchronisation. Use of: pressure gauge, synchroniser tool, tachometer, oscilloscope, diagnostic tester, flow meter, vacuum gauge. Comparing results to manufacturer's figures, dismantling, checking and rebuilding procedures.	**MSM07, MSM15**		
6 hours	**Single and multi-point injection systems**: Operation of injection systems Airflow sensor, fuel supply system, fuel pump, filter, fuel regulator, injectors, sequential injection, continuous injection, semi-continuous injection, electronic control unit (ECU) and injector pulse width, sensors, throttle bodies, slide throttles, barrel throttles. Operation of each system under various operating conditions including cold starting, warm up, hot starting, acceleration, deceleration, cruising and full load. Explain engine speed limiting and knock sensing.	**MSM07**		
6 hours	**Engine management systems:** Function and purpose of engine management systems. Open loop and closed loop control, types of input and output devices. Digital components and systems. Closed loop system, integrated ignition and injection systems, operation under various conditions. Analogue, digital, programmable and non-programmable systems.	**MSM07**		

The specific and appropriate Health & Safety precautions and legislative requirements should be integrated into the delivery of each aspect of engine technology to ensure relevance and meaning to Candidates. The legislation and generic provisions and are outlined in units G1, G2 & G3.

Hours	Topic	Technical Certificate Unit Mapping	On-line Assessment	Practical Assessment
3 hours	**Engine Mechanical Symptoms and Faults:** Symptoms and faults related to worn cylinders, cylinder liners, pistons, piston rings, crankshaft, camshaft, bearings, cylinder head and gasket, valves, valve seats and valve guides, cam-belts; lubrication system and components.	MSM07		
3 hours	**Assessment, repair and restoration of mechanical engine components:** Assessing engine mechanical components. Measuring for wear and serviceability including cylinder bores, cylinder heads and gasket, crankshaft journals, valve faces; valve guides, valve seats, camshafts, bearings, cam belts, lubrication system components, turbocharger, supercharger. Methods used for the repair and restoration of engine components.	MSM07		
6 hours	**Diagnosis & testing of motorsport engine mechanical systems:** Locate and interpret information for diagnostic tests. Preparation of tools and equipment for use in diagnostic testing. Conduct systematic assessment of engine components and systems, component condition, engine balance, power balance, performance and operation, wear, run out, alignment; use of appropriate tools and equipment including compression gauges, leakage testers, cylinder balance tester, pressure gauges, micrometers and vernier gauges. Evaluation of test results from diagnostic testing. Comparison of test result and values with vehicle manufacturers' specifications and settings. Procedures for dismantling components and systems. Assess, examine and measure components. Identification of probable faults, malfunctions, incorrect settings and wear. Rectification or replacement procedures. Evaluation of repair to confirm system performance.	MSM07		

(Continued)

The specific and appropriate Health & Safety precautions and legislative requirements should be integrated into the delivery of each aspect of engine technology to ensure relevance and meaning to Candidates. The legislation and generic provisions and are outlined in units G1, G2 & G3.

Hours	Topic	Technical Certificate Unit Mapping	On-line Assessment	Practical Assessment
1.5 hours	**Ignition system failure or malfunctions:** No spark, misfiring, backfiring, cold or hot starting problems, poor performance, pre-ignition, detonation, exhaust emission levels, fuel consumption, low power and unstable idle speed.	MSM07		
1.5 hours	**Petrol and diesel injection system failures or malfunctions:** Cold or hot starting problems, poor performance, exhaust emissions, high fuel consumption, erratic running, low power and unstable idle speed.	MSM07		
1.5 hours	**Engine management system failure or malfunctions:** Misfiring, backfiring, cold or hot starting problems, poor performance, pre-ignition, detonation, exhaust emission levels, fuel consumption, low power and unstable idle speed.	MSM07		
7 hours	**Diagnosis and testing of engine auxiliary systems:** Locating and interpreting information for diagnostic tests manufacturer's, vehicle and equipment specifications. Use of equipment, testing procedures, test plans, fault codes and legal requirements. Preparation of tools and equipment for diagnostic testing. Conducting systematic assessment of systems including: component condition and performance, component settings, component values, electrical and electronic values; system performance and operation, use of appropriate tools and equipment including gauges, multimeter, breakout box, oscilloscope, diagnostic tester, manufacturer's dedicated equipment, exhaust gas analyser, fuel flow meter and pressure gauges. Evaluation and interpretation of test results. Comparison of test results, values and fault codes with vehicle manufacturer's specifications and settings. Procedures for dismantling, components and systems using appropriate equipment. Explain the methods of assessing, examining and measuring components including: settings, input and output values, voltages, current consumption, resistance, output patterns with oscilloscope, condition, wear and performance of components and systems.	MSM07		

The specific and appropriate Health & Safety precautions and legislative requirements should be integrated into the delivery of each aspect of engine technology to ensure relevance and meaning to Candidates. The legislation and generic provisions and are outlined in units G1, G2 & G3.

Hours	Topic	Technical Certificate Unit Mapping	On-line Assessment	Practical Assessment
	Identification of probable faults and malfunctions, incorrect settings, wear, values, inputs and outputs and fault codes. Rectification or replacement procedures. Evaluation of components and systems following repair to confirm system performance.			
Total Hours 80.5	**These hours do not include covering Level 1 and 2 requirements**			

ADVANCED TRANSMISSION TECHNOLOGY

The specific and appropriate Health & Safety precautions and legislative requirements should be integrated into the delivery of each aspect of transmission technology to ensure relevance and meaning to Candidates. The legislation and generic provisions and are outlined in units G1, G2 & G3.

Hours	Topic	Technical Certificate Unit Mapping	On-line Assessment	Practical Assessment
6 hours	**Electronic Technology for Transmission Systems:** Operation of electrical and electronic systems and components related to light vehicle transmission systems including ECU, sensors and actuators, electrical inputs & outputs, voltages, oscilloscope patterns, digital and fibre optic principles. Interaction between the electrical/electronic system with hydraulic system and mechanical components. Electronic and electrical safety procedures.	**MSM13, MSM15**		
8 hours	**Automatic Transmission Systems:** Fluid couplings, fluid flywheel, torque converter (torque multiplication, efficiency) Operation of epicyclic gearing (sun, planet, annulus and carrier) and method for achieving different gear ratios. Hydraulic control systems, components and operation. Electronic control system, components and operation. Continuously variable transmissions (CVT). Sequential manual gearbox (SMG).	**MSM13, MSM15**		

(Continued)

The specific and appropriate Health & Safety precautions and legislative requirements should be integrated into the delivery of each aspect of transmission technology to ensure relevance and meaning to Candidates. The legislation and generic provisions and are outlined in units G1, G2 & G3.

Hours	Topic	Technical Certificate Unit Mapping	On-line Assessment	Practical Assessment
6 hours	**Driveline Systems:** Limited slip differential. 4 wheel drive systems including third differential, differential locks. Ferguson, Torsen, torque split. Traction control systems and launch control. Competition propeller and drive shafts – CV joints, materials, strength and balance.	MSM13, MSM15		
4.5 hours	**Competition Clutches:** Single plate, paddle, organic, ceramic, sintered, multi-plate, cable operated, hydraulic, basket clutch, release bearing, upgraded pressure plates.	MSM15		
4.5 hours	**Competition Gearboxes:** Synchro, dog, ratios for circuit/venue/engine, changing gear clusters, strengthening, oil and coolers, materials and finishes.	MSM15		
3 hours	**Transmission Symptoms and Faults:** Clutch and coupling faults: abnormal noises, vibrations, fluid leaks, slip, judder, grab, failure to release. Manual Gearbox, abnormal noises, vibrations, loss of drive, difficulty engaging or disengaging gears. Automatic gearbox, abnormal noises, vibrations, loss of drive, failure to engage gear, failure to disengage gear, leaks, failure to operate, incorrect shift patterns, electrical and electronic faults. Final drive, abnormal noises, vibrations, loss of drive, oil leaks, failure to operate, electrical and electronic faults. Drivelines and couplings, abnormal noises, vibrations, loss of drive.	MSM13		

The specific and appropriate Health & Safety precautions and legislative requirements should be integrated into the delivery of each aspect of transmission technology to ensure relevance and meaning to Candidates. The legislation and generic provisions and are outlined in units G1, G2 & G3.

Hours	Topic	Technical Certificate Unit Mapping	On-line Assessment	Practical Assessment
7 hours	**Diagnosis and Testing of Transmission System:** Location and interpretation of information for diagnostic tests, vehicle and equipment specifications, use of equipment, testing procedures, test plans, fault codes, legal requirements. Prepare equipment for use in diagnostic testing. Systematic testing and inspection of transmission system, mechanical, hydraulic, electrical and electronic systems using appropriate tools and equipment including, multimeters, oscilloscope, pressure gauge. Workshop based and road testing of transmission system. Evaluation and interpretation of test results from diagnostic testing and/or road testing. Comparison of test result and values with vehicle manufacturers' specifications and settings. Dismantle components and systems using appropriate equipment and procedures. Assessing, examining and evaluation of operation, settings, values, condition and performance of components and systems.	MSM13		
	Probable faults, malfunctions, incorrect setting. Rectification or replacement procedures. Evaluation of systems to confirm operation and performance.	MSM13		
Total Hours 39	**These hours do not include covering Level 1 & 2 requirements**			

ADVANCED ELECTRICAL & ELECTRONIC TECHNOLOGY

The specific and appropriate Health & Safety precautions and legislative requirements should be integrated into the delivery of each aspect of electrical technology to ensure relevance and meaning to Candidates. The legislation and generic provisions and are outlined in units G1, G2 & G3.

Hours	Topic	Technical Certificate Unit Mapping	On-line Assessment	Practical Assessment
4.5 hours	**Advanced lighting technology:** Xenon lighting, gas discharge lighting, ballast system, LED, intelligent front lighting, blue lights, complex reflectors, fibre optic, optical patterning. **Lighting systems:** Circuits, sidelights, dipped beam, main beam, dim/dip, interior lights, map lights, spot lamps, rain lamps, fault diagnosis for lighting systems,	MSM15		
4.5 hours	**Motorsport electric components:** Operation and construction of motors, relays, interfaces, modules, switches, batteries, starter systems, charging systems, circuit protection, cable specification and termination.	MSM15		
1.5 hours	**Sensors:** Temperature, pressure, movement.	MSM15		
1.5 hours	**Spark plugs:** Type, range, reach.	MSM15		
3 hours	**Electronic control systems and mapping for:** Ignition, injection, turbo control, launch control, traction control, gearboxes, and differentials.	MSM15		
Total Hours 15	**Does not include covering Level 1 & 2 requirements**			

GENERAL ASPECTS

The specific and relevant Health & Safety precautions and legislative requirements should be integrated into the delivery of the aspects vehicle and engine technology wherever possible to ensure relevance and meaning to Candidates, the generic aspects of health and safety are shown below. Only an outline of the main requirements of the Acts and Regulations is required and an awareness of how the legislation relates to the work of trainees. It is recommended that these units are delivered in a Motorsport specific context as appropriate whilst maintaining there generality related to the motor industry for future possible employment. If the candidate has achieved the Level 2 Technical Certificate they do not need to revisit these units – G1, G2 and G3 – at Level 3.

Hours	Topic	Technical Certificate Unit Mapping	On-line Assessment	Practical Assessment
6 hours	**Main Health & Safety Legislation:** Outline main provisions and legal duties imposed by HASAWA, COSHH, EPA, Manual Handling Operations Regulations 1992, PPE Regulations 1992. **Regulations specific to job role of trainees:** Awareness of: Health & Safety (Display Screen Equipment) Regulations 1992 Health & Safety (First Aid) Regulations 1981 Health & Safety (Safety Signs and Signals) Regulations 1996 Health & Safety (Consultation with Employees) Regulations 1996 Employers Liability (Compulsory Insurance) Act 1969 Confined Spaces Regulations 1997 Noise at Work Regulations 1989 Electricity at Work Regulations 1989 Electricity (Safety) Regulations 1994 Fire Precautions Act 1971 Reporting of Injuries, Diseases & Dangerous Occurrences Regulations 1985 Pressure Systems Safety Regulations 2000 Waste Management 1991 Dangerous Substances and Explosive Atmospheres Regulations (DSEAR) 2002 Control of Asbestos at Work Regulations 2002 **Legislative requirements for use of work equipment:** Provision and Use of Work Equipment Regulations 1992 Power Presses Regulations 1992 Pressure Systems & Transportable Gas Containers Regs 1989 Electricity at Work Regulations 1989 Noise at Work Regulations 1989 Manual Handling Operations Regulations 1992 Health and Safety (Display Screen Equipment) Regulations 1992 Abrasive Wheel Regulations Safe Working Loads **Workplace policies and procedures.**	**G1, G2, MSM07, MSM08, MSM13**		

(Continued)

The specific and relevant Health & Safety precautions and legislative requirements should be integrated into the delivery of the aspects vehicle and engine technology wherever possible to ensure relevance and meaning to Candidates, the generic aspects of health and safety are shown below. Only an outline of the main requirements of the Acts and Regulations is required and an awareness of how the legislation relates to the work of trainees. It is recommended that these units are delivered in a Motorsport specific context as appropriate whilst maintaining there generality related to the motor industry for future possible employment. If the candidate has achieved the Level 2 Technical Certificate they do not need to revisit these units – G1, G2 and G3 – at Level 3.

Hours	Topic	Technical Certificate Unit Mapping	On-line Assessment	Practical Assessment
6 hours	**Organisational requirements for maintenance of the workplace:** Company Health & Safety Policy, Health & Safety Executive. Trainee's personal responsibilities and limits of their authority with regard to work equipment. Risk assessment of the workplace activities and work equipment. Person responsible for training and maintenance of workplace equipment. When and why safety equipment must be used. Location of safety equipment. Hazards associated with their work area and equipment. Prohibited areas. Plant and machinery that trainees must not use or operate. Why and how faults on unsafe equipment should be reported. Storing tools, equipment and products safely and appropriately. Using the correct PPE. Manufacturers' recommendations. Location of routine maintenance information e.g. electrical safety check log.	**G1, G2, MSM07, MSM08, MSM13**		
6 hours	**Safely use tools and equipment:** Files, saws, hammers, screwdrivers, drill and drill bits, spanners, punches, measuring equipment, air tools, taps and dies, vices and sockets. Pillar/bench drills, abrasive wheels, presses and lead lights. Axle stands, hydraulic jacks, vehicle lifts/hoists and engine cranes/hoists. Air tools. Welding equipment, power cleaning equipment and exhaust gas extraction. Waste disposal and cleaning equipment. Brake testing equipment. Tool storage and accessibility.	**G1, MSM07, MSM08, MSM13**		

The specific and relevant Health & Safety precautions and legislative requirements should be integrated into the delivery of the aspects vehicle and engine technology wherever possible to ensure relevance and meaning to Candidates, the generic aspects of health and safety are shown below. Only an outline of the main requirements of the Acts and Regulations is required and an awareness of how the legislation relates to the work of trainees. It is recommended that these units are delivered in a Motorsport specific context as appropriate whilst maintaining there generality related to the motor industry for future possible employment. If the candidate has achieved the Level 2 Technical Certificate they do not need to revisit these units – G1, G2 and G3 – at Level 3.

Hours	Topic	Technical Certificate Unit Mapping	On-line Assessment	Practical Assessment
4.5 hours	**Basic maintenance procedures for equipment:** Hand tools. Electrical equipment. Mechanical equipment. Pneumatic equipment. Hydraulic equipment. **Requirement to clean tools and work area.** Requirement to carry out the housekeeping activities safely and to minimise inconvenience to customers and staff. **Risks** involved when using solvents and detergents. Storage and disposal of waste, used materials and debris correctly. **Inspection of equipment to identify faults:** Hand tools. Electrical equipment. Mechanical equipment. Pneumatic equipment. Hydraulic equipment. Reporting unserviceable tools and equipment.	**G1, MSM07, MSM08, MSM13**		
	Economical use of resources: Heating, electricity, water, consumables, other energy sources.	**G1**		
	Requirement to clean tools and work area: Maintenance procedures and methodologies, good and safe housekeeping, risks involved with solvents and detergents. **Disposal of waste materials:** Safe system of work, storage and disposal of waste materials, regulations and requirements to dispose of waste, used materials and debris correctly, advantages of recycling.	**G1, MSM07, MSM08, MSM13**		

(Continued)

The specific and relevant Health & Safety precautions and legislative requirements should be integrated into the delivery of the aspects vehicle and engine technology wherever possible to ensure relevance and meaning to Candidates, the generic aspects of health and safety are shown below. Only an outline of the main requirements of the Acts and Regulations is required and an awareness of how the legislation relates to the work of trainees. It is recommended that these units are delivered in a Motorsport specific context as appropriate whilst maintaining there generality related to the motor industry for future possible employment. If the candidate has achieved the Level 2 Technical Certificate they do not need to revisit these units – G1, G2 and G3 – at Level 3.

Hours	Topic	Technical Certificate Unit Mapping	On-line Assessment	Practical Assessment
3 hours	**Requirements when driving vehicles:** Legal requirements when using vehicles on the road, road safety requirements, lighting, tyres, steering, braking, seat belts, road worthiness. **Legal requirements for the driver and the vehicle**: Appropriate driver's licence, road fund licence, vehicle insurance, MOT regulations. Requirements when driving vehicles (company owned, customers) on the road: seat belts, speed limits, care of vehicle, adherence to **Highway Code**. Requirements of the Road Traffic Act. Race Licence Requirements. MSA Regulations.	**G1, G2, G3, MSM07, MSM08, MSM13**		
4.5 hours	**Health & Safety Requirements of Vehicle Repair:** Vehicle protection and personal protection (PPE) when working on vehicles. **Hazards and risks** involved in repair removal and replacement of units and systems; safety precautions and procedures involved with mechanical, electrical and electronic repair or dismantling. Requirements for disposal of old units, materials, components and fluids. **Fire hazards and safety**: Fire extinguishers, actions in the event of a fire, fire drill and fire exits. **Dealing with accidents at work** – procedures. **Personal conduct** in vehicle workshop situations: awareness and care of others and avoidance of inappropriate behaviour.	**G2, G3, MSM07, MSM08, MSM13**		

The specific and relevant Health & Safety precautions and legislative requirements should be integrated into the delivery of the aspects vehicle and engine technology wherever possible to ensure relevance and meaning to Candidates, the generic aspects of health and safety are shown below. Only an outline of the main requirements of the Acts and Regulations is required and an awareness of how the legislation relates to the work of trainees. It is recommended that these units are delivered in a Motorsport specific context as appropriate whilst maintaining there generality related to the motor industry for future possible employment. If the candidate has achieved the Level 2 Technical Certificate they do not need to revisit these units – G1, G2 and G3 – at Level 3.

Hours	Topic	Technical Certificate Unit Mapping	On-line Assessment	Practical Assessment
6 hours	**Technical information relating to vehicle repair:** **Sources technical & repair information:** Vehicle specifications, identification codes, service schedules, MOT testing requirements, equipment information, procedures for use of equipment repair procedures and test plans. **Types of information:** Paper based, hard copy manuals, computer stored data, on-board diagnostic displays, CD ROM, internet, manufacturer's website. Documentation involved in vehicle repair and maintenance processes: company job cards, manufacturer's service schedules, test plans, inspection sheets, MOT requirements, customer requirements and in-vehicle service record. **Types of Communication:** Verbal, written, and electronic. Communications involved in vehicle repair, signs and notices, memos, telephone, electronic mail, vehicle job card, informed and uninformed people. Written word, spoken word, symbolic gestures, visual images, multimedia. **Costs:** Relationship between time, costs and profit. **Economical use of resources**: Heating, electricity, water, consumable materials e.g. grease. **Reporting delays** and/or additional work required to relevant supervisory person, referral of problems, additional work.	**G2, G3, MSM07, MSM08, MSM13**		

(Continued)

The specific and relevant Health & Safety precautions and legislative requirements should be integrated into the delivery of the aspects vehicle and engine technology wherever possible to ensure relevance and meaning to Candidates, the generic aspects of health and safety are shown below. Only an outline of the main requirements of the Acts and Regulations is required and an awareness of how the legislation relates to the work of trainees. It is recommended that these units are delivered in a Motorsport specific context as appropriate whilst maintaining there generality related to the motor industry for future possible employment. If the candidate has achieved the Level 2 Technical Certificate they do not need to revisit these units – G1, G2 and G3 – at Level 3.

Hours	Topic	Technical Certificate Unit Mapping	On-line Assessment	Practical Assessment
1.5 hours	Organisation of Dealerships & Vehicle Repair Organisations: Function of main sections: reception, body shop, repair workshop, paint shop, valeting, parts department, administration office, vehicle sales. Interrelationships of departments. Organisational structure. Job Roles.	**G3**		
1.5 hours	**Developing positive working relationships:** Importance of positive working relationships with regard to: morale, productivity, company image, customer relationships. **Reasons and effects.** **Listening to the views of others.** **Honouring commitments.**	**G3**		
6 hours	**Motorsport Organisation: the organisation and structure of Motorsport including:** FIA, MSA, international, national and local clubs. Scrutineering, marshalling, Clerk of Course, circuits and race organisers. **Introduction to the special needs of Motorsport specific customers and their vehicles.**	**Not specified as a G Unit; but recommended to maintain context**		
Total Hours 45				

INDEX

Index

Printed and bound by CPI Group (UK) Ltd, Croydon, CR0 4YY
17/10/2024
01775697-0011